GREEN AND SUSTAINABLE BUSINESS PRACTICE

An ISEB Foundation Guide

BCS The Chartered Institute for IT

Our mission as BCS, The Chartered Institute for IT, is to enable the information society. We promote wider social and economic progress through the advancement of information technology science and practice. We bring together industry, academics, practitioners and government to share knowledge, promote new thinking, inform the design of new curricula, shape public policy and inform the public.

Our vision is to be a world-class organisation for IT. Our 70,000 strong membership includes practitioners, businesses, academics and students in the UK and internationally. We deliver a range of professional development tools for practitioners and employees. A leading IT qualification body, we offer a range of widely recognised professional qualifications.

Joining BCS The Chartered Institute for IT

BCS qualifications, products and services are designed with your career plans in mind.

We not only provide essential recognition through professional qualifications but also offer many other useful benefits to our members at every level.

Membership of BCS demonstrates your commitment to both your own and the IT community's professional development. We continue to invest heavily in developing and delivering effective member services that IT practitioners value and use.
www.bcs.org/membership

Further Information

BCS The Chartered Institute for IT, First Floor, Block D, North Star House, North Star Avenue, Swindon, SN2 1FA, United Kingdom.
T +44 (0) 1793 417 424
www.bcs.org/contact

GREEN IT FOR SUSTAINABLE BUSINESS PRACTICE
An ISEB Foundation Guide

Mark G. O'Neill

bcs
The Chartered Institute for IT

Published by British Informatics Society Limited (BISL), a wholly owned subsidiary of BCS The Chartered Institute for IT, First Floor, Block D, North Star House, North Star Avenue, Swindon, SN2 1FA, UK. www.bcs.org

ISBN 978-1-906124-62-5

British Cataloguing in Publication Data.
A CIP catalogue record for this book is available at the British Library.

Typeset by Lapiz Digital Services, Chennai, India.
Printed at CPI Antony Rowe, Chippenham, UK.

CONTENTS

LIST OF FIGURES AND TABLES

AUTHOR

Mark G. O'Neill is a highly experienced IT Professional with 20 years experience in the IT industry. He has held a number of senior roles and has experience of large-scale IT Service Management and Infrastructure implementation programmes.

Mark's experience in customer relations and the managed services arena enabled him to successfully manage delivery of IT service provision for the Sales and Marketing division of one of Europe's largest organisations. More recently Mark has been appointed the Head of Service Management and Green Learning Consultancy for QA Ltd, and in the recent past has undertaken a number of key consultancy assignments and has extensive training experience with leading organisations in the public and private sector, both in the UK and overseas.

In conjunction with the BCS, Mark has also been responsible for designing the first formally accredited training course for Green IT – the Foundation Certificate in Green IT, and is the holder of the world's first individual accreditation in Green IT.

FOREWORD

Increasing worldwide concern for the environment has brought into sharp focus the emissions and energy inefficiencies of IT systems. As the evidence for man-made climate change mounts, and we are all faced with having to do more with fewer resources, there is an increasing demand for authoritative sources of practical information to help people use their IT more efficiently. Across a range of activities the mantras of 'switch off' and 'sweat the asset' are effective first steps; however, there are greater gains to be had where these can be harnessed as part of a wider programme of efficiency savings and strategic planning which can deliver more significant and sustained savings across the whole of an organisation's IT estate, e.g. through joint IT and Estates adoption of the EU Code of Conduct for energy efficient data centres. And IT itself is also part of the solution, enabling emission and energy savings in other areas of an organisation's operations, e.g. by providing services that support remote working and electronic meetings, savings can be made on travel, paper and accommodation.

In this book, Mark O'Neill provides strategies and practical approaches to resolving the environmental problems that IT poses, thereby enabling savings for the benefit of all concerned. He sets out a path to embedding Green IT in sustainable business practices that will help organisations avoid the charge of 'Green Wash' and achieve real energy, emission and bottom line financial savings of interest to us all as we face up to our current economic and climate change challenges.

Bob Crooks MBE, CITP
Defra Lead for Green IT
Chair of the BCS Green IT Specialist Group

ACKNOWLEDGEMENTS

Many parties have contributed to the development and production of this book, and without their help this book would never have been written. I would like to start by acknowledging my colleagues at QA Ltd, including Mark Willis, Alan Morgan, Geraldine Lee and Jeff Payne. Particular thanks go to both Mark and Alan for their guidance and editing and proofing skills, Geraldine for her contribution in working with me to develop the Green IT and ITIL® concepts and Jeff for his continued support and championing of Green IT. I would also like to express my gratitude to Sarah O'Brien at the Green Electronics Council for her contribution with regard to EPEAT® and Dr Michael Gell for his contribution relating to the UK's Carbon Reduction Commitment. Thanks also to Gary Mills from Decorum Technology for his input into intelligent homes and building technology and Jacek Truszczynski for giving me permission to include detailed information relating to ENERGY STAR®. Also my gratitude goes to Dave Berwick from the Corin Group for his input regarding virtualisation and the reuse of existing technology.

I would also like thank my family and friends for their patience and understanding whilst I was writing this book, and, in particular, my wife Sam and children Shannon and Aidan. Finally I would like to thank all the other contributors who I have not named personally, but without whom this book would never have seen the light of day.

1 INTRODUCTION TO GREEN IT

From both an individual and an organisational perspective the global environmental issues we all face are here to stay for the foreseeable future. Arguably the most commonly known and widely reported environmental concern nowadays is global climate change. The Earth's climate is influenced mainly by the amount of heat being produced by the sun but other factors, including the production of GreenHouse Gases (GHGs) in the Earth's atmosphere, and the properties of the Earth's surface, determine how much of this heat is retained or reflected back to space. The general consensus amongst scientists and climate specialists is that the amounts of man-made GHG produced, such as carbon dioxide (CO_2), have increased significantly since the beginning of the Industrial Revolution, which started in the UK in the mid 1800s. This is mainly due to the burning of fossil fuels, land use change, intensive farming and modern agricultural working practices. Although the scientific evidence linking climate change to severe weather events such as floods, rising sea levels and droughts has recently been challenged,[1] organisations still need to consider the social implications of their working practices as many of the severe weather events tend to affect the poorest of people and nations, many of whom are unable to defend themselves against such events.

For many of us in the Information Communication and Technology (ICT) Industry, the environmental spotlight is shining directly at us, as the consumption of energy derived from traditional fossil fuels becomes an international concern. It is widely cited in Information Technology (IT) circles that the ICT industry is responsible for approximately 2 per cent of worldwide carbon emissions, which is roughly the equivalent to the carbon emissions attributed to the aviation industry. Whilst this figure is open to interpretation and observers would question whether the current situation requires any great attention, what is indisputable is that IT is in a unique position to influence the other 98 per cent.

The demand for technology services from both an individual and an organisational perspective is increasing rapidly. This demand is being generated from an increased reliance on ICT to provide solutions for both our business and personal challenges. This includes, but is not restricted to, the increased use of electronic transactions in financial services such as online banking and electronic trading, the growing use of the internet for social interaction, communication and entertainment, the move to having electronic medical records for health care, the growth in global commerce and services, and the adoption of satellite navigation for both personal and organisational use.

This effectively means that organisations large and small are going to have to make a concerted effort to operate and function very differently from the way they do currently if they are to directly influence the issues highlighted above. Whether or not we decide to believe the science, there will be a time in the not-too-distant future when organisations and, indeed, individuals will have no choice but to adhere to legislation and governance related to reduction of CO_2 and other GHGs.

Green IT for Sustainable Business Practice sets out to provide guidance to IT service providers who want to improve their standing in the eyes of their users and customers. The book is also designed to be used by organisations who wish to understand the way in which IT can be asked to support the organisation's Green agenda. By helping the organisation to realise its environmental aspirations, and by making a recognised contribution to the reduction of the organisation's GHG emissions, IT service providers can use Green IT as a means to ensure that IT is seen as a force of good against evil in a world in which a large number of cases perceives IT as more of a hindrance than an enabler; this can only be a good thing.

Regardless of the environmental issues with which organisations have to contend, there are many other factors driving us to adopt best practice, especially in the ICT industry. Therefore, the commitment needed from the ICT industry to reduce carbon emissions can be achieved using two distinct approaches.

First, there is an opportunity to identify and highlight the areas of ICT that are directly contributing to an organisation's GHG emissions, and to recommend solutions to reduce both the primary and secondary emissions associated with the delivery of ICT into the organisation. Recent studies have shown that, globally, the electricity consumption of Personal Computers (PCs) is growing by 5 per cent year on year. In an average Small-to-Medium-sized Enterprise (SME) electricity consumption accounts for 10 per cent of an IT department's budget, alarmingly rising to over 50 per cent in some extreme cases. Recent studies have also found that the cost of electricity to run a typical computer over its lifetime is now even greater than its purchase price. Perhaps the most worrying consideration is that between 2000 and 2005, the amount of electricity consumed by data centres worldwide doubled – and it was estimated in 2007 by the US Environmental Protection Agency (EPA) that by the end of 2011, in the United States alone, 10 new power stations will need to have been built to cope with this ever-growing demand.

Second, organisations will need to appreciate that investment in IT can actually help to reduce the emissions associated with the organisation as a whole, even if this means that the carbon emissions associated with ICT increase. An example of this is where organisations are investing in intelligent building-monitoring tools that require investment in the ICT infrastructure. The investment in the new infrastructure may lead to the carbon emissions associated with the IT department increasing; however, it will also lead to a decrease in the organisation's overall carbon footprint.

The Carbon Trust describes a carbon footprint as 'the total GHG emissions caused directly and indirectly by a person, organisation, event or product'.[2] The footprint calculation needs to include all six of the Kyoto Protocol recognised GHGs: CO_2, methane (CH_4), nitrous oxide (N_2O), hydrofluorocarbons (HFCs), perfluorocarbons (PFCs) and sulphur hexafluoride (SF_6) and is measured in tonnes of CO_2 equivalent (tCO_2e). The CO_2 equivalent allows the different GHGs to be compared on a like-for-like basis, relative to one unit of carbon. CO_2e is calculated by multiplying the emissions of each of the six GHGs by its 100-year global-warming potential.

The overall carbon footprint is made up of emissions from all the activities across the organisation. These can include direct emissions that result from activities that the organisation controls and its energy usage, and indirect emissions from products and services that the organisation does not directly control. The investment in ICT in order to reduce the organisation's overall carbon footprint is sometimes referred to as 'silicon trading'.

A key intention of this book is to help organisations and individuals to understand the critically important political, financial, social and legal drivers surrounding Green IT. As with many 'Green' initiatives, the perception is that if the term 'Green' is used, then it is going to cost money and require substantial investment. Fortunately, in the majority of cases, it is actually the complete opposite. Green IT, as with any best-practice programme, is a common-sense approach to delivering cleaner, less expensive ICT infrastructure which will deliver both environmental and financial efficiencies across the organisation and assist the organisation in lowering its overall carbon footprint.

2 THE DEFINITION OF GREEN IT

Within the ICT industry there are many commentators who cannot decide whether Green IT is the correct description for what is a relatively new concept. But whatever the arguments with regard to terminology, there is a set of clear definitions when describing the subject matter.

Green IT is a collection of strategic and tactical initiatives that directly reduces the carbon footprint of an organisation's computing operation. This can manifest itself in many different forms, the vast majority of which are discussed in this book. However, Green IT is not just focused on reducing the environmental impact of the ICT industry. It is also focused on using the services of ICT to help reduce the organisation's overall carbon footprint, regardless of the type, shape or size of the organisation.

There are few individuals or organisations that would argue that IT or telecommunications do not impact on their lives. In fact, it is hard to think of one organisation or set of circumstances where IT does not have any influence. Therefore, the potential for IT to deliver Greener goods, services or individual lifestyles is enormous. Traditionally, in many organisations IT is perceived (rightly or wrongly) as a barrier to progress, when in reality it should be the complete opposite. It is no exaggeration to state that IT and the related telecommunications have revolutionised the way in which businesses operate and individuals lead their lives.

A prime example of this is the way the retail banking industry now operates compared to the pre-internet age. Before the internet was widely available and trusted, the majority of consumers performed their personal banking on the high street. In reality (for those of us old enough to remember), it meant having to make a journey to the nearest branch office, standing in a queue with numerous other bank customers and waiting to be seen by a bank employee before discussing your banking needs. Nowadays, the majority of personal banking is performed over the internet, and the banking industry as a whole has been revolutionised.

However, does this provide a Greener solution? As with any debate, there are arguments to be made on both sides. The Green IT lobby will point to fewer car journeys being made to the branch offices by the bank's customers, and a reduced requirement to heat and light buildings, all of which are perfect examples of a Greener way of working. However, others will point to the fact that to deliver the online banking solution, huge amounts of IT infrastructure are needed, which leads not only to increased power consumption, but also to increased embodied emissions.

Another example of IT enabling a business to dramatically change the way it operates is the loyalty card schemes now adopted by the majority of supermarkets and high-street stores. The loyalty card scheme has revolutionised the manner in which supermarkets manage their supply chain. By being able to collect, collate and manipulate extensive data on their customers' spending habits, the supermarkets are able accurately to predict demand patterns and order their stock accordingly. This leads to less waste being sent to landfill and therefore fewer associated GHG emissions. Once again this is a business or industry sector using its IT capabilities to deliver more efficient and effective means of operating.

Green IT also encourages and supports Greener behaviour by the organisation's employees, customers and suppliers. By various means, including awareness campaigns and ongoing education, and in some cases legislation, the whole culture of an organisation can be changed.

However, it is also widely accepted that for a cultural change to take place in an organisation, there has to be a visible commitment from the top downwards. For the organisation's executive body to only report and be driven by its financial performance is no longer good enough or acceptable. Therefore, more and more organisations are now making key strategic decisions based on their Triple Bottom Line (TBL).

TBL Accounting attempts to link the social and environmental impact of an organisation's activities, in a measurable way, to its economic performance in order to show improvement, or to make evaluation more in-depth. In practical terms, TBL accounting means expanding the traditional reporting framework to take into account ecological and social performance in addition to financial performance. This is more commonly referred to as people, profit and planet. With the introduction of TBL accounting, organisations will be acutely aware of how their own working practices will have an effect on the associated working practices of their subsidiaries and suppliers. For example, the organisation will not tolerate the use of child labour and would commit to paying fair and equitable salaries to its employees, would maintain a safe work environment and tolerable working hours, and would not otherwise exploit a community or its labour force. If the organisation is not directly responsible for working practices, it may decide to support schemes such as Fairtrade.[3] A TBL business also typically seeks to reinvest a proportion of its wealth by contributing to the strength and growth of its community with health care and education schemes.

Of course, successful cultural change can start from the 'bottom up' by implementing quick and easy initiatives and by driving through a collective focus of sustainability across the whole organisation. For example, simply by encouraging one or two individuals to switch off their PCs before they go home at night, a 'chain reaction' can be set off within the organisation which will lead to other individuals and eventually whole departments doing likewise. Once the hearts and minds of the individuals have been won, organisations will quickly experience a collective change taking place.

There will, however, always be the pessimists who will doubt whether the culture of an organisation can really be changed, never mind in a relatively short space of time. In recent times and in many countries, we have seen examples of a change in culture that very few people would have predicted a few years previously, including a ban on smoking in public places. Of course, legislation was a key driver in implementing this change in the way people lead their lives and perhaps it can be argued that without a change in the law that made smoking in public places an offence, it would never have happened. However, in many countries now, Green-related legislation is already in place and new environmental laws such as the Climate Change Act in the UK will be the driver to cultural change. In April 2010, the UK Carbon Reduction Commitment (CRC), the UK's first mandatory carbon-trading scheme, came into effect, aiming to reduce the current level of carbon by 1.2 million tonnes of CO_2 per year by 2020. The scheme is expected to affect around 5,000 UK organisations, in both the public and the private sector.

Finally, by implementing a Green IT policy, organisations can ensure the sustainability of the resources used by IT. Later in the book, we investigate ways in which we can reduce, reuse and recycle infrastructure. In essence, this will lessen the environmental impact of continually upgrading and replacing the organisation's software and hardware.

3   ESTABLISHING A GREEN IT POLICY

The first step for any IT organisation wanting to improve its Green credentials is to establish a Green IT policy. The Green IT policy will explain to staff, suppliers, customers and users where the organisation stands on Green and environmental issues related to ICT. In order for it to be credible, it must be adopted, authorised and committed to by the organisation's senior management.

To avoid the possibility of duplication or of IT being accused of doing its own thing, the Green IT policy needs to be aligned to any organisational environmental strategies that may already exist. The most probable owner of any existing strategy to which the Green IT policy can align will be the person who is accountable for Corporate and Social Responsibility (CSR) in the organisation. Once the person or organisational department responsible for CSR has been identified, checks need to be made and any possibility of duplication has to be identified. The organisation should capitalise on this opportunity to develop products and services that will establish Green IT as best practice.

Establishing a Green IT policy is not necessarily going to be easy to achieve. Currently, the vast majority of organisations are facing tough economic challenges and the establishment of a Green IT policy may not be high up on the list of most organisations' priorities. However, there is a real possibility that a Green IT policy will deliver a substantial reduction in costs as well as carbon emissions. A Green IT policy will also assist organisations to align to their relative carbon reduction targets, and therefore reduce the risk of incurring fines for poor environmental performance.

For many organisations' shareholders and board members, achieving the organisation's carbon reduction targets and saving money do not go hand in hand. The general perception is that Green initiatives, IT or otherwise, can be costly, and a return on investment in Green initiatives will not be realised for many years. Therefore, the organisation's Green IT policy needs to contain initiatives that will convince all the stakeholders that it will deliver reduced costs, cut emissions and improve the organisation's environmental credentials.

To ensure that the Green IT policy will receive senior management and stakeholder sponsorship and support, three key steps of accountability need to be recognised. The first key step is reporting the cost. The Green IT policy must be accounted for in the organisation's overall IT budget, or environmental strategy financial planning. IT directors and Chief Information Officers (CIOs) who present to the board and stakeholders a well-written business case demonstrating the financial benefits and cost savings of Green IT will more than likely find their ideas well received and support from the board forthcoming. However, organisations must not be afraid to 'think big' or include long-term, high-investment initiatives in their Green IT policy. As long as the costs and a Return On Investment (ROI) have been identified in the supporting business case, there is no reason not to include them.

The next step to achieving successful 'buy in' for the Green IT policy is to ensure that it has a set of structured stages. First, identify the 'quick wins'. These will be the high-impact, low-cost or low-investment initiatives. Begin with an initial burst of cost-effective, highly visible activities that deliver the quick wins, for example, a poster campaign asking your users to switch off their PCs at night.[4] It is generally recognised that the financial decision-makers in most organisations are averse to taking risks. This is especially true in a recession or if the organisation is experiencing financial challenges. Therefore, the decision-makers need to have hard evidence of how a Green IT policy will work in their organisation and the financial rewards it will bring. However, the rewards are not just restricted to financial recompense: there are other key areas to consider, including environmental, social and legal drivers. The Green IT policy should start with articulating simple, easily applied changes, which might include an awareness campaign and double sided printing as a default printer setting, before moving on to longer term high-investment projects such as server virtualisation.

Last, your Green IT policy needs a Green IT sponsor. This person needs to be someone in the organisation who is not only senior, but also feels passionately about the initiative, and is willing to represent Green IT at board level. Potentially, Green IT could influence every element of the business and could impact across every organisational context, from the factory floor to the boardroom. Accountability needs to be sought at the highest level as this will send a clear message that Green IT is part of the corporate strategy, and the organisation's employees and shareholders must take it seriously.

Moving forward, Green IT and the Green IT policy will go beyond the organisation. It will encourage and enthuse employees to demand greater sustainability from suppliers and partners. It will drive the need for longer warranty periods which will lead to a slowing down of hardware replacement and refresh projects, and manufacturers will be encouraged to design systems and infrastructure that will be easier to upgrade and repair. Already, we are seeing a shift in ethos amongst mobile-phone manufacturers who recently agreed to provide universal mobile phone chargers,[5] thus helping to reduce electronic waste that traditionally ends up in landfill.

Having a Green IT policy will encourage organisations to consider calculating their carbon footprint, and implementing initiatives to reduce it. This in turn will encourage manufacturers to improve the 'Greenness' of their products, reducing emissions and expenditure throughout the supply chain.

Green IT leaders need to think about how they can ensure that the environmental impact of their ICT practices remains high on the list of the organisation's priorities despite pressures to reduce costs. The key is to link environmental strategy to money-saving objectives, and to lead the change at boardroom level. It is now no exaggeration to state that there has never been a more compelling set of circumstances to ensure that organisations take Green IT seriously. Organisations are facing a 'perfect storm' where the forces of legislation, ecology, economics and publicity collide and it is all too easy to see Green IT as being just about the environment, but, as has already been stated, that is just one aspect. Organisations also need to significantly reduce their running costs, be ready to meet carbon emission and energy reduction targets and fend off the bad publicity and shareholder backlash that comes from not being environmentally responsible. Without a Green IT policy, organisations cannot hope to address these risks, opportunities and challenges.

4 GREEN WASH

The term 'Green Wash' is used to describe the practice of misrepresenting products or services as eco-friendly, for either personal or organisational gain, when in reality the opposite is true. For example, manufacturers may claim that their electrical or electronic product is Green because it consumes less power than that of a rival product. However, it may very well be that the product has in fact been manufactured using hazardous materials and it may actually contain an enormous embodied carbon footprint. Unfortunately, the manufacturer will neglect to pass on this information to its customers and clients, thereby deceiving them into thinking that they are an eco-friendly product manufacturer.

Another practice carried out by large or multi-national organisations is to promote one or two of its sector's Green credentials, whilst hiding away or failing to report on the organisation as a whole. This is an accusation made on a regular basis against the large supermarket chains. At the forefront of their environmental campaigns will be highly visible and relatively inexpensive working practices such as giving a penny back to the customer for each plastic bag they reuse. However, what is not made clear or reported on is the initiatives that will reduce carbon emissions relating to unnecessary food packaging or their end-to-end transport supply chain, both of which could potentially be far more harmful to the environment in the longer term.

Another practice that can be described as Green Wash is where an organisation will attempt to achieve carbon reduction targets without really putting pragmatic carbon reduction initiatives in place. An example of this was reported in January 2010.[6] In 2008 the UK Government ordered large energy companies to invest in measures for improving energy efficiency and cutting fuel poverty. Companies were able to choose how to meet their obligations, with each measure funded, and were given a score for the lifetime carbon savings that that would achieve. As a means of meeting these measures, one large energy company posted 12 million low-energy light bulbs to households across the UK as part of its legal obligation to cut carbon emissions, despite government advice that many would never be used.

The company was facing a fine of more than £40 million, or 10 per cent of its annual turnover, if it failed to meet its target for improving efficiency in homes under the carbon emissions reduction target scheme. To prevent having to pay the fine, the energy company saved millions of pounds by giving away the bulbs rather than implementing alternative ways of meeting its obligation, such as insulating homes, which was considered to be much more effective but nearly seven times more expensive.

The UK Department of Energy and Climate Change admitted in June 2009 that the scheme was flawed and resulted in significant wastage. The department said:

Government is increasingly concerned that the number of lamps already distributed has been so high that it may work out at more than the average number of highest-use light fittings in a house. As such, there is an increasing risk to carbon savings under the scheme where lamps are not used, are installed on low-use light fittings, or replace existing low-energy bulbs.

It said that direct mail-outs of bulbs would be banned from 1 January 2010, allowing six months for companies to wind down their schemes.

The energy company began posting 12 million bulbs in November 2009, five months after the ban had been announced and just as the postal system was struggling to cope with the increased volume of mail in the run up to Christmas. A spokesperson for the energy company said that the scheme was designed to be completed on New Year's Eve, hours before the ban came into force at midnight. The spokesperson admitted that the company did not know how many of the bulbs would be used.

There is nothing under the carbon emissions reduction target scheme that means we have to get evidence that bulbs are being used. It's up to the customer.

Companies were allowed to register immediate carbon savings from every bulb issued on the assumption that all recipients instantly installed them in some of their most intensively used light sockets. In reality, many people either stored the bulbs or threw them away, often because they were the wrong fitting or wattage. To compound the issue, hundreds of thousands of households had received more than 180 million free or subsidised low-energy bulbs in the previous eight months and a survey in July 2009 by the Energy Saving Trust found that the average home had at least six unused bulbs lying in drawers and cupboards.

The energy companies of course can point to the fact that they are legally doing nothing wrong and simply meeting their targets – in the highly competitive industry sectors in which they work, it is hardly surprising that they would opt for the least costly option. According to the latest government estimates, each low-energy bulb costs an energy company £2.97 and saves 0.04 tonnes of carbon over its lifetime, whereas insulating the external solid walls of a three-bedroom semi-detached house costs £8,760 and saves 18.08 tonnes of carbon. A company can achieve the same score of 18.08 tonnes by posting 452 bulbs, costing only £1,342. In the first 18 months of the scheme, companies issued 182 million bulbs but insulated only 17,000 solid-wall homes. Britain has 6.6 million solid-wall homes requiring insulation. Companies can pass on all the costs of the scheme to their customers. Over three years it is expected to add more than £100 to the average household's energy bills.

As well as being aware of Green Wash campaigns, organisations also need to be alerted to the dangers of malicious individuals and products intent on hijacking the Green IT bandwagon. One recent example of this is a virus specifically aimed at organisations and individuals who are committed to Green IT working practices, called GreenAV, also referred to as Green AV, Green Antivirus 2009 or GreenAntivirus2009. It is a rogue anti-spyware application that is designed to coerce PC users into believing that their computer system has been infected with various spyware and malware, all of which are tactics employed to ensure that the unsuspecting user will fall for the scam and ultimately purchase the full version of GreenAV, which (unsurprisingly) is completely useless.

GreenAV usually infects computers via misleading advertisements, which claim to be online virus scanners. Alternatively, GreenAV may also enter into a computer system via Trojan infections. Once GreenAV is installed on a computer it will perform a system security scan and will report malware infections found, all of which are fake. It will also display pop-up warning messages and system alert notifications, stating that the computer is infected. These warnings are also fictitious. Not only is the software dangerous, it is ethically corrupt stating that if purchased for a one-off fee of US$100, US$2 will be donated to saving the Amazon rain forest. Clearly this is a false statement with no factual evidence that any donations whatsoever are being made.

Nevertheless, there are positive signals that organisations and, in particular, those companies that are predominantly in the public eye are now taking the practice of Green Wash seriously. In the last quarter of 2006, the UK Advertising Standards Authority (ASA) received 62 complaints relating to Green Wash advertising, which grew to 268 complaints in the first six months of 2007. However, there were only 264 for the whole of 2008 and only 158 by November 2009.

The assumption is that this recent decline in complaints demonstrates that organisations are finally realising the risk of huge political and economic damage to their organisation's reputation, if they are found to be deceiving their customers, shareholders and the public in general.

5   KEY ROLES IN GREEN IT

Any organisation committed to implementing policies relating to Green IT will need to ensure that it identifies the roles that need to be fulfilled. This will include process owners, communication managers, implementation teams (both technical and non-technical) and all other key stakeholders. A stakeholder is anyone who will be impacted, involved, or even just consulted in the Green IT initiative.

Once the stakeholders have been identified, the next step is to map the elements of the initiative of which they need to have visibility. Table 5.1 is an example of how this may look.

Table 5.1 Green IT stakeholder identification for information share

Stakeholders	Strategic direction	Financial reporting	Operational changes	Competitive position
Public shareholders	X	X		X
Trade unions	X		X	
Service desk	X		X	
Press and media	X	X		X
Executive board	X	X	X	X
Regulatory bodies		X		
Procurement manager	X		X	
CIO	X	X	X	X

THE GREEN IT CHAMPION

The purpose of the Green IT Champion is to support and encourage the organisation's stakeholders, including third parties and suppliers, to ensure that all Green IT issues are (wherever possible) identified, managed and ultimately resolved, and in such a way that the organisation meets its own environmental and sustainability objectives.

The Green IT Champion needs to ensure that the benefits of Green IT are communicated wherever possible. This needs to be achieved by taking ownership of the promoting, developing and delivering of the organisation's key Green IT messages via an agreed communication plan. The Green IT Champion must also encourage colleagues and partners to adhere to Green IT initiatives and to act as the focal point for Green IT, promoting a range of Green IT activities. There will also be a requirement to facilitate and chair regular meetings and workshops for other individuals involved in or interested in Green IT. The key attribute of the Green IT Champion is enthusiasm about the subject matter and complete commitment to its delivery. The role can be either paid or voluntary.

Specific responsibilities will include to:

- develop and deliver a high-profile launch for Green IT across the organisation;
- build positive working relations with all the stakeholders;
- develop a Green IT focus group and identify sources of information to assist in enabling the project;
- develop a project delivery plan, communications plan and marketing materials;
- identify and assess relevant legislation and governance for Green IT, for example, the Waste Electrical and Electronic Equipment (WEEE) directive;
- compile a risk register identifying where non-compliance of legislation is likely, and identify and document all relevant risk-reduction measures;
- if required, recruit and secure resources to carry out organisational and departmental audits and prepare and train them for carrying out the audits;
- develop or procure a Green IT audit tool;
- complete departmental Green IT environmental audits to identify specific areas of poor performance;
- develop an action plan for each departmental area to provide solutions for the areas of poor performance identified in the audits;
- agree sign-off of the plans and secure funding;
- complete a benchmark behavioural survey of staff and stakeholders and manage the emotional cycle of change before making any significant changes to working practices;
- produce the narrative and images for a Green IT web page, launch it and ensure that it is kept up to date;
- provide detailed comparative feedback to participating departments measured against agreed benchmarks;

- liaise with the organisation's marketing department to produce exciting and informative media stories that promote the Green IT achievements implemented in the organisation;

- manage the project budget and ensure timely delivery of key financial objectives;

- manage project administration, maintain project records and produce management reports as required;

- represent the project and organisation at key Green IT and sustainable management events;

- develop a cost benefit analysis model to calculate the estimated date when the initial investment in Green IT is paid off and when Green IT starts to become profitable to the organisation;

- support other organisational environmental initiatives in conjunction with the Corporate and Social Responsibility Unit.

THE CHIEF INFORMATION OFFICER (CIO)

The CIO is accountable for all IT functions of the organisation and for directing the information and data integrity of the organisation. This includes being responsible for all data centres, technical service centres, production scheduling functions, Service Desks, communication networks (voice and data), computer program development and computer systems operations.

The purpose of the CIO is to set the strategic direction of all information processing and communication systems and operations, and to provide overall management and definition of all computer and communication activities within the organisation. Given the nature and importance of the role, the CIO has a vital part to play in ensuring that Green IT and sustainable working practices are considered in every context. The CIO must be seen as supporting and providing sponsorship of the Green IT policy and setting the strategic direction for the rest of the organisation.

The CIO must provide leadership as well as direction once the various operational and tactical Green IT initiatives start to take place. The CIO is also responsible for interaction with the executive management team to monitor and validate the organisation's compliance with its environmental policies, which includes (but is not limited to) legislation such as the CRC.

Specific responsibilities will include to:

- define, update and implement IT strategy including Green IT and be accountable for Green IT across the organisation;

- align IT objectives and programmes to the organisation's environmental and carbon reduction objectives and strategies;

- control performance objectives and overall IT budget, including Green IT expenditure;

- provide evidence in the form of management reporting with defined metrics of the Green IT initiatives being achieved and how they support the overall environmental objectives of the business;

- select, manage and control IT providers including managed outsourced service providers and maximise leverage with third parties to provide environmentally friendly and sustainable services;

- consolidate IT across the organisation and reduce the costs of services, in conjunction with the delivery of reduced carbon and other GHG emissions.

CORPORATE SOCIAL RESPONSIBILITY (CSR) MANAGER

Responsible for an organisation's CSR policy, the CSR manager ensures that the organisation has a built-in, self-regulating mechanism to monitor adherence to the key drivers of legislative, political, social and environmental requirements and ethical standards.

The purpose of the CSR manager is to ensure that organisations adopt and comply with best corporate responsibility practice and commit the organisation to act responsibly and to take its environmental and social responsibilities seriously.

Specific responsibilities will include to:

- integrate and steer strategic direction of CSR;

- ensure compliance and management of key issues to drive continuous improvement – adding both organisational and customer value;

- raise education and awareness of CSR within the business to enable staff to become informed company ambassadors whose actions, decisions and behaviours are driven by company values;

- appoint a CSR committee to shape, develop and implement a CSR strategy to integrate CSR in the business and to deliver value;

- ensure that strategy is aligned to and enhances both corporate values and the organisation's brand;

- develop and implement policies and procedures to support strategic direction and ensure that CSR dovetails into other business activities including Green IT;

- work collaboratively with stakeholders to ensure management of key priority CSR issues, threats and opportunities;

- review, recommend and oversee development of an Environmental Management System (EMS) ensuring appropriate resource and business integration, resulting in a clear and communicable position on the organisation's approach to reducing its carbon emissions;

- develop and implement a cohesive community investment strategy that includes all community impact areas from donations, staff fund-raising and volunteering to education liaison, economic development and cause-related marketing;

- introduce an action-plan framework and performance-tracking system to monitor the associated activities and to drive continuous improvement;
- develop internal processes and systems to capture key CSR data;
- recommend and be responsible for benchmarking and auditing current working practices relating to the role of CSR;
- identify and leverage involvement with regional and national CSR partners and specialist interest groups, and promote and improve the business's reputation for CSR;
- establish a clear and communicable position on the organisation's CSR both internally and externally;
- ensure that channels are maximised to engage with both staff and customers on CSR to understand and fulfil needs and interests;
- develop and implement CSR communications plan to increase staff education and awareness, articulate what it means for them and the role they play in the organisation's CSR commitment;
- develop and implement CSR communications plan to customers and other key stakeholders including shareholders and third-party suppliers;
- raise the profile of CSR overall, so that the organisation is seen as an innovator in CSR.

CHIEF SUSTAINABILITY OFFICER (CSO)

The role of the Chief Sustainability Officer (CSO) is a relatively new one that has grown from the more traditional role of the organisation's Health and Safety Executive. This evolution has taken place because of the desire, and in some cases absolute need, to understand the organisation's commitment to its environmental and Green programmes.

The purpose of the CSO is to represent the organisation's needs and desires relating to sustainability and the environment, balanced with delivering key business objectives such as the organisation's profitability and growth.

Specific responsibilities will include to:

- interact directly with senior personnel at regulatory agencies, advisory groups and associations aimed at influencing public policy developments and regulations relating to the environment;
- provide sponsorship and representation at board level of the organisation's sustainability and environmental strategies;
- proactively identify and manage specific challenges to the organisation's value chain from either legislative or non-legislative directives;
- implement and manage the organisation's commitment to reducing its carbon footprint;

- share the responsibility with other chief executives for research and development of new technologies and strategies to deliver Green IT;
- ensure regular interaction with the chief executive officer, the board of directors and sub-committees to develop innovative projects for new ventures to deliver the sustainability goals of the organisation;
- combine sustainability, environmental and health and safety responsibilities to provide overall protection for the organisation;
- be accountable at a senior level for environmental and sustainability strategies and auditing and monitoring programmes;
- understand and be accountable for the delivery of countermeasures against strict regulations, both domestically and internationally;
- be accountable for and manage the reduction of GHG emissions, dangerous working practices, non-environmentally friendly product design, waste handling and other harmful environmental and health and safety practices.

PROCUREMENT MANAGER

An organisation's procurement manager will be responsible for buying goods and services in line with relevant environmental directives and the organisation's own Green IT policy. The procurement manager will also identify and demonstrate carbon savings achieved through negotiation of supplier contracts whilst developing relationships with key suppliers and will encourage them to improve their own organisation's Green credentials.

The purpose of the procurement manager is to be accountable for supporting cost-effective procurement of products and services whilst achieving or exceeding customer and business needs, including Green IT and other environmental objectives.

Specific responsibilities will include to:

- implement the organisation's strategic and tactical procurement plans whilst improving the purchasing capabilities across all organisational functions, including Green IT;
- work with cross-functional team members to achieve team objectives in supplier/market analysis;
- utilise relevant procurement practices and tools such as the Electronic Product Environmental Assessment Tool (EPEAT) to ensure environmentally friendly procurement of IT infrastructure;
- compile market research data, inclusive of environmental analysis for sustainable purchasing;
- support the procurement process by conducting environmental supplier evaluations;
- prepare for negotiations involving critical or sensitive suppliers;

- be responsible for the implementation of initiatives to ensure that environmentally friendly products, services and processes are supported by quality performance metrics;

- resolve day-to-day supply chain and supplier issues.

Of course, the dynamics, size and maturity of organisations will differ enormously from one organisation to another. Therefore, in small-to-medium-sized businesses, you may well find that some of these roles are combined. Conversely, in larger organisations there may well be more than one CSR or procurement manager and the responsibilities will be shared.

Either way, for the Green IT policy implementation to be successful, shared understanding and clearly defined lines of responsibility and accountability need to be identified. Therefore, it is often worth taking the time to think through the roles that organisations need and to identify the critical tasks that must be managed. Without this clarity, you will most likely find gaps, duplication and confusion. Teamwork will be disconnected and inefficient and you are less likely to deliver the desired results.

In these situations, the delegation of tasks and other responsibilities can be too important to leave to chance. Therefore, the utilisation of a methodology that will help to reduce this risk is imperative. One of the most popular methodologies employed for this type of exercise is the **RACI matrix.** The RACI matrix is a responsibility assignment matrix system that brings structure and clarity to assigning the roles that people play within a team. RACI is an acronym for Responsible, Accountable, Consulted and Informed. Their definitions are:

Responsible
The individuals who perform the tasks or activities, and are responsible for the action or implementation of the task. The degree of responsibility is defined by whoever is Accountable. There is typically one role with a participation type of Responsible, although others can be delegated to assist in the work if required.

Accountable
The individuals who are ultimately accountable for the correct and final completion of the deliverable or task, and the one to whom Responsible is accountable. In other words, an Accountable must sign off (approve) the work that Responsible delivers. There must be only one Accountable specified for each task or deliverable.

Consulted
Those whose opinions are sought; and with whom there is two-way communication.

Informed
Those who are kept up-to-date on progress, often only on completion of the task or deliverable, and with whom there is just one-way communication.

A simple grid system can be used to ensure that all tasks have been identified and assigned. Table 5.2 is an example of how the RACI matrix can be applied to Green IT.

Table 5.2 How the RACI matrix can be applied to Green IT

Green IT activities	Green IT Champion	CSO	IT user	Procurement manager
Develop Green IT policy	R	A	I	C
Facilitate awareness workshops	R A	I C	I	I
Evaluate supplier contracts to ensure WEEE compliance	C	I	I	R A
Ensure that PCs are switched off at the end of the day	R A	R	R	RC
Own and improve Green IT policy	R	A R	I C	I C

6   KEY FACTORS DRIVING GREEN IT

The key factors driving Green IT can be identified in four specific categories: Political, Environmental, Social and Legal.

POLITICAL

Politically, if an organisation chooses not to take Green IT seriously, then the potential fallout could be extremely damaging. With the internet now influencing every aspect and element of our everyday lives, politically minded organisations, competitors or even individuals can cause untold damage to the reputation of the organisation. Via websites, blogs, podcasts and newsgroups shareholders, investors and customers can be encouraged to boycott the targeted organisations if it is felt that their products or working practices are either ethically or environmentally unacceptable.

One of the most high-profile organisations in recent years to feel the wrath of both the media and the Green campaigners was oil and gas giant Shell. Through a combination of both Green Wash and environmentally unfriendly working practices, Shell found that large parts of its operation were being intensely scrutinised and reported on. The oil company was accused of making misleading claims about its action to tackle climate change whilst withdrawing investments from renewable energy supplies.

On 13 August 2008, the UK's Advertising Standards Authority upheld a complaint from environmental campaigners the World Wide Fund for Nature (WWF) that an advertisement placed by Shell in the Financial Times suggested that oil sands were a sustainable energy source. The ASA – the independent body responsible for regulating UK advertising – branded the advert 'misleading', due to its ambiguous use of the word 'sustainable'.

The advert referred specifically to the company's oil sands deposits in Alberta, Canada, and their work to build the largest oil refinery in North America in Port Arthur, Texas. The WWF complained that Shell's repeated use of the term 'sustainable' was entirely at odds with these activities. A subsequent report released by the WWF and Co-operative Financial Services revealed that the production of oil from tar-soaked shale or sand can create up to eight times as many emissions as conventional oil production does. David Norman, Director of Campaigns at WWF-UK, said:

> Oil sands are one of the world's dirtiest sources of fuel and have a major impact on the environment. Their extraction cannot be described as a sustainable process and for Shell to claim otherwise is wholly misleading.

Half of the remaining arboreal forest in the world is situated in Canada and large areas of this have already been destroyed by oil-sand extraction. Alberta is also home to some of the largest dam structures in the world that were built to hold in huge tailing ponds of waste water. These ponds, some of which are visible from space, are the toxic by-product of the oil sand industry. David Norman stated:

> The ASA's decision to uphold WWF's complaint sends a strong signal to business and industry that Green wash is unacceptable ... Oil sands are an incredibly destructive source of energy and, along with the expansion of Shell's oil refining capacity in Texas, cannot be considered a sustainable way to meet the world's future energy needs. If Shell were serious about sourcing sustainable energy, then they would be far better placed investing in renewable energy, such as wind, tidal or solar power.[7]

As the focus of media and environmental pressure groups widens, how long before the ICT industry starts to come under increased scrutiny and media pressure? Organisations must be ready to be more accountable about their working practices and be prepared to be open and honest about their Green credentials if they are to prevent political and inevitably economic damage to their organisation's reputation.

ENVIRONMENTAL

Some business activities have an enormous capacity to damage the life support systems of Earth and the ICT sector is no different. As we have already identified, the ICT industry also has the potential to make an enormous contribution to the reduction of CO_2 and other GHGs. In its press release in June 2008, The Climate Group stated the following:[8]

> Smarter technology and transformation in the way people and businesses use technology could reduce annual man-made global emissions by 15 per cent by 2020 and deliver energy efficiency savings to global businesses of over EUR 500 billion (GBP 400 billion/US$ 800 billion). In its report, *SMART 2020: enabling the low carbon economy in the information age*, international management consultants McKinsey & Company, stated that whilst ICT's own sector footprint will almost double by 2020, ICT's unique ability to monitor and maximise energy efficiency both within and outside of its own sector could cut CO_2 emissions by up to five times this amount. This represents a saving of 7.8 Gt CO_2 equivalent by 2020.

Although teleworking, videoconferencing, epaper and ecommerce are increasingly commonplace, the report notes that replacing physical products and services with their virtual equivalents (dematerialisation and substitution) is only one part (6 per cent) of the estimated low-carbon benefits the ICT sector can deliver. Far greater opportunities for emissions savings exist in applying ICT to global infrastructure and industry and the McKinsey & Company report examines four major opportunities where ICT can make further transformational cuts in global emissions. These exist globally within smart building design and use, smart logistics, smart electricity grids and smart industrial motor systems.

Steve Howard, CEO, for The Climate Group, said:

> PCs, mobile phones, and the web have transformed the way we all live and do business. Global warming and soaring energy prices mean that rethinking how every home and business uses technology to cut unnecessary costs and carbon is critical to our environment and economy. Supported by innovative government policy, ICT can unlock the clean Green industrial revolution we need to tackle climate change and usher in a new era of low carbon prosperity.

Luis Neves, Chair of the Global e-Sustainability Initiative (GeSI), said:

> The ICT industry is a key driver of low carbon growth and can lead transformation towards a low carbon economy and society. The ICT sector must act quickly to demonstrate what is possible, require clear messages from policy makers about targets and continue to radically innovate to reduce emissions.

Achim Steiner, UN Under-Secretary General and Executive Director, UN Environment Programme (UNEP), said:

> This rigorous assessment underlines that the world can realise a Green economy and make the transition to a low carbon economy. It also underlines the crucial importance of the international community reaching a deal on a new climate agreement at the climate convention meeting in Copenhagen in 2009. This partnership between GeSI, convened under UNEP, The Climate Group and McKinsey gives us yet another platform for action and yet another compelling reason for reasoned optimism.

Key Findings of the SMART 2020 Report

(i)   The global ICT footprint. A new 'socially networked' generation around the world continues to drive unprecedented global demand for ICT hardware, software and services providing mobile and instant access to information. A global study predicts that PC ownership will quadruple between 2007 and 2020 to four billion devices, and emissions will double over the same period, with laptops overtaking desktops as the main source of global ICT emissions (22 per cent); mobile phone ownership will almost double to nearly five billion by 2020 but emissions will only grow by 4 per cent; and broadband uptake will treble to almost 900 million accounts over the same period, with emissions doubling over the entire telecoms infrastructure. Despite the major anticipated advances in the energy efficiency of products, the ICT sector's own footprint – currently 2 per cent of global emissions – is expected to grow at 6 per cent per year compound annual growth rate (CAGR) and double by 2020, driven by increased technology uptake in India, China and the rest of the world. Trends such as virtualisation of data centres, long-life devices, smart chargers, next-generation networks and growth of renewable energy consumption (e.g. solar-powered base stations) could help to deliver future sustainable sector growth. To help, rather than hinder, the fight against climate change, the ICT sector must manage its own growing impact and continue to reduce emissions from data centres, telecommunications networks and the manufacture and use of its products.

(ii) ICT's enabling effect in cutting global emissions. Crucially, the SMART 2020 Report articulates significant opportunities for emissions reductions and how cost savings can be leveraged by applying ICT to global infrastructure and industry. If global businesses systematically used ICT to realise all of the solutions indicated in the report, they would unlock global energy efficiency savings of over EUR 500 billion (calculated as at December 2007). This enabling effect is due to ICT's unique ability to measure, optimise and therefore manage energy consumption. Four major global opportunities were examined through regional case studies:

(a) Applied globally, smart motors and industrial automation would reduce 0.97 Gt CO_2 emissions in 2020, worth EUR 68 billion (US\$ 107.2 billion). A review of manufacturing in China uncovered that, without technological improvements, 10 per cent of China's emissions (2 per cent of global emissions) in 2020 will come from China's motor systems alone: to improve China's industrial efficiency by even 10 per cent would deliver up to 200 million tonnes of CO_2 emissions savings.

(b) The global emissions savings from smart logistics in 2020 would reach 1.52 Gt CO_2 emissions with energy savings worth 208 billion euro (US\$ 441.7 billion). In Europe, the logistics industry looks set to grow by 23 per cent between 2002 and 2020. Through a host of efficiencies in transport and storage, smart logistics in Europe could deliver fuel, electricity and heating savings of 225 Mt CO_2 emissions in 2020.

(c) Buildings are the second highest consumer of power in the world behind industry. Globally, smart buildings technologies would enable 1.68 Gt CO_2 emissions savings, worth 216 billion euro (US\$ 340.8 billion). A closer look at buildings in North America indicates that through better building design, management and automation, 15 per cent of North America's building emissions could be avoided.

(d) Smart Grid technologies were the largest opportunity explored in the study, and could globally reduce 2.03 Gt CO_2 emissions worth 79 billion euro (US$ 124.6 billion). In India, currently over 30 per cent of the generated power is lost through Aggregated Technical and Commercial losses (AT&Cs). Reducing these losses in India's power sector by 30 per cent is possible through better monitoring and management of electricity grids, first with Smart meters and then through integrating more advanced ICTs into the so-called 'energy internet'.

(iii) Getting Smart about ICT going forward. The report recommends that a Smart framework is implemented, outlining key actions required by the ICT sector, national governments and industry. Transformation of the economy will occur when Standardisation (S), Monitoring (M) and Accounting (A) of energy consumption prompt a Rethink (R) in how we optimise for energy efficiency and how we live, work and play in a low-carbon world. Through this enabling platform, Transformation (T) will occur when the business models that drive low-carbon alternatives can be developed and diffused at scale across all sectors of the economy.

SOCIAL

More and more organisations are now realising the need to meet the social expectations of their employees or prospective employees and to provide evidence that their organisations are supporting and delivering ethical and environmental working practices. Green policies are now commonplace in most organisations' mission statements. The Sarbanes Oxley Act (USA, 2002), which regulates corporate governance, lays out not only responsible money-accounting practices but also an environmental accounting of company impacts and plans for sustainable development. A recent US poll indicated that social responsibility is highly valued by American consumers, and that damaging the environment is the main reason for consumers to think a company socially irresponsible, brands will not be able to opt out of being Green. The poll also found that American consumers between the ages of 18 and 29 are more likely to buy organic, environmentally friendly or Fair Trade products than other age groups.

In an address to the World Economic Forum on 31 January 1999, the former Secretary-General of the United Nations, Kofi Annan, challenged business leaders to join an international initiative, the Global Compact, that encourages organisations to work together with UN agencies and labour and civil societies to support universal environmental and social principles.[9] The Global Compact's operational phase was launched at UN Headquarters in New York on 26 July 2000.

Today, thousands of organisations globally are working to advance 10 universal principles. The 10 principles are encompassed in 4 categories: human rights, labour, environment and anti-corruption. The UN Global Compact's environment principles are derived from the Rio Declaration on Environment and Development, which contains a pledge to 'undertake initiatives to promote greater environmental responsibility; and encourage the development and diffusion of environmentally friendly technologies'. Refer to Table 6.1.

Table 6.1 The UN Global Compact's 10 principles

Core values	Principles
	Businesses should:
Human rights	(i) support and respect the protection of internationally proclaimed human rights;
	(ii) make sure that they are not complicit in human rights abuse;
Labour standards	(i) uphold the freedom of association and the effective recognition of the right to collective bargaining;
	(ii) eliminate all forms of forced and compulsory labour;
	(iii) effectively abolish child labour;
	(iv) eliminate discrimination in respect of employment and occupation;
Environment	(i) support a precautionary approach to environmental challenges;
	(ii) undertake initiatives to promote greater environmental responsibility;
	(iii) encourage the development and diffusion of environmentally friendly technologies;
Anti-corruption	(i) work against corruption in all its forms, including extortion and bribery.

LEGAL

Whether or not organisations choose to embrace the political, environmental or social drivers of Green IT, in the future they may not have any choice. Both the ICT service provider and the organisation as a whole will have to comply with globally impacting acts of law, as new and challenging legislation is brought into being. New laws and corporate reporting requirements have been outlined in the Climate Change Act 2008, which is aimed at reducing the UK's carbon emissions by 2050 by 80 per cent, measured against 1990 levels.

The Climate Change Act 2008 makes the UK the first country in the world to have a legally binding long-term framework to cut carbon emissions. It also creates a framework for building the UK's ability to adapt to climate change. The Climate Change Act finished its passage through Parliament on 18 November 2008, and was enacted by Royal Assent on 26 November 2008.

The Act:

- requires the UK Government to publish five-yearly carbon budgets;
- creates a Committee on Climate Change;
- requires the Committee on Climate Change to advise the Government on the levels of carbon budgets to be set, the balance between domestic emissions reductions and the use of carbon credits and whether the 2050 target should be increased;
- places a duty on the Government to assess the risk to the UK from the impacts of climate change;
- provides powers to establish trading schemes for the purpose of limiting GHG;
- confers powers to create waste-reduction pilot schemes;
- amends the provisions of the (UK) Energy Act 2004 on renewable transport fuel obligations.

It is probably the last three requirements outlined above that will impact both ICT and non-ICT organisations, especially those who rely on large data centres and distributed infrastructure to deliver their computing requirements. However, once again ICT has an opportunity to either directly or indirectly assist organisations in meeting their CO_2 emission reduction targets.

Key areas for ICT to explore can include reduced reliance on traditional fossil fuel by investing in renewable energy sources such as solar, wind and tidal energy generation. An example of this is the intention of Google to build floating data centres. The data centres will use wave and water movement to provide power, and the sea air to provide natural cooling. Google's application for a 'Water-based data centre' patent was filed in February of 2007 and published in 2009. It describes a floating platform-mounted computer data centre comprising a plurality of computing units, a sea-based electrical generator in electrical connection with the plurality of computing units and one or more sea-water cooling units for providing cooling to the plurality of computing units.

The majority of the patent deals with the logistics of ship-based data centres, although it also examines the use of wave power, tidal power and sea water for providing electricity and cooling to land-based data centres that are close enough to water. An example of a land-based but close to water data centre is the Mauritius Eco-Park, which plans to develop a system to use Sea Water Air Conditioning (SWAC) to support data-centre tenants. The data centre will be reliant on the deep water currents that bring colder water to within two miles of Mauritius.

The Eco-Park plans to build a system of pipes that will extend two miles offshore and as much as 1,000 m (3,200 ft) beneath the ocean surface, where the water is approximately 5 °C (40 °F). The cold water will be piped back to the data centre complex and used in the facility's cooling system, eliminating the need for power-hungry cooling and air conditioning systems, as demonstrated in Figure 6.1.

27

Figure 6.1 Sea Water Air Conditioning (SWAC) data centre technology

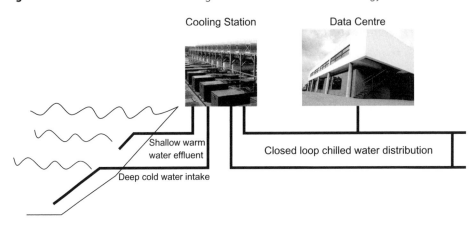

Another sector where ICT can have a major impact is in technology that provides improved and more efficient energy use in buildings, which subsequently delivers reduced carbon emissions.

The field of intelligent buildings, intelligent homes and Building Management Systems (BMSs) encompasses an enormous variety of technologies, across commercial, industrial, institutional and domestic buildings, including energy management systems and building controls. The purpose of 'intelligent buildings' concepts is to control, monitor and optimise building services. From an environmental perspective, the predominant focus will be on lighting and heating.[10]

The potential to reduce carbon emissions within these concepts and the surrounding technology is vast and any facilities or building managers considering premises development or site relocation should also consider the opportunities presented by intelligent buildings technologies and concepts.

The essence of BMSs and intelligent buildings is in the control technologies, which allow integration, automation and optimisation of all the services and equipment that provide services and manage the environment of the building concerned. The use of these technologies allows the optimisation of various site and building services.

There are numerous methods by which building services within buildings can be controlled, falling broadly into two method types: time-based – providing heating or lighting services etc. only when required, and optimiser parameter-based – often utilising a representative aspect of the service, such as temperature for space heating or luminance for lighting.

Time-based controls can be used to turn on and off the heating system (and/or water heating) at pre-selected periods (of the day, of the week etc.). Optimiser parameters make sure that the building reaches the desired temperature when occupancy starts and provides protection against freezing or frost protection. Compensated systems will control flow temperature in the heating circuit relative to external temperature. This will give a rise in the circuit flow temperature when the outside temperature drops.

Lighting control methods are time-based control and optimiser parameter-based methods where a level of luminance or particular use of lighting is required. Used pragmatically, lights are switched on corresponding to the use and layout of the lit areas in order to avoid lighting a large area if only a small part of it needs light. Also, technology can be exploited to switch on and off automatically in each zone to a preset schedule for light use.

Passive Infra-Red (PIR) occupancy sensing can be utilised in areas which are occupied intermittently, for example, server rooms and remote data centres. The occupancy sensors are used to indicate whether or not anybody is present and switch the light on or off accordingly.

Until recent years, energy efficiency has been a relatively low priority and low perceived opportunity to building owners and investors. However, with the dramatic increase and awareness of environmental energy use concerns and government legislation, the advances in cost-effective technologies to provide energy efficiency is fast becoming part of facilities management and operations strategy. The concepts are also now making significant inroads into the domestic residential house-building sectors.

For lighting, energy savings can be up to 75 per cent of the original circuit load, which represents 5 per cent of the total energy consumption of the residential and commercial sectors. Energy savings potential from water heating, cooling or hot water production can be up to 10 per cent, which represents up to 7 per cent of the total energy consumption of the domestic residential and commercial sectors. Experience from studies in Austria suggests that potential heating and cooling energy savings are up to 30 per cent in public buildings. Even allowing for the fact that buildings used in the study may have been those with particularly high energy usage, the figure is a compelling one.[11]

In association with environmental legislation and standards, Health and Safety Regulations and global trends towards improving indoor air quality standards are also significant drivers of the need for BMSs and intelligent buildings.

7 BUSINESS AND ORGANISATIONAL BENEFITS OF ADOPTING GREEN IT

By adopting Green IT the business or organisation can derive many benefits, but predominantly they will be realised through improved sustainable operations, improved organisational reputation, a shift towards a pragmatic culture within the organisation and, of course, reduced cost.

It is a critical success factor of the Green IT project to understand and promote the business benefits of taking an environmentally friendly approach towards IT service provision and this must be at the heart of any related business case. Green IT has the capability of reducing the Total Cost of Ownership (TCO) of IT for the organisation, enhancing the reputation of the organisation with its customers and shareholders, and installing a culture of enterprise, social responsibility and a 'can do' attitude within the organisation.

We have already documented the four key factors of Green IT (Political, Environmental, Social, Legal) and these will be at the forefront of an organisation's willingness to promote itself as a conscientious and modern-day thinking organisation. Many organisations are now seeking to have their Green credentials recognised publicly. This can be done either formally by benchmarking and auditing against recognised standards such as ISO 14001 or by other less formal means such as media surveys and business sector awards.[12]

One such initiative is *The Sunday Times* 60 Best Green Companies competition, which aims to identify the country's most environmentally aware companies with the most environmentally engaged workforces.[13] To determine the best Green companies, both the employer and the employee are surveyed.

The 2009 employer survey covered environmental management policies, environmental training and internal consultation, energy consumption, waste production and recycling. It also considered efforts to promote environmental initiatives within business supply chains.

The employee survey put 52 statements to staff which they rated from 'strongly agree' to 'strongly disagree'. These statements fell into four broad areas: policies and systems, training and motivation, reporting and communications and environmental performance. They included statements on workplace practices, for example, 'my organisation always does what is best for the environment' and 'my organisation always goes for the cheapest, rather than the most environmentally friendly option'. With regard to management attitudes, statements include 'my boss encourages me to think about energy saving'; and for personal opinions and behaviour, 'environmental training I receive at work makes me think differently about what I do at home' and 'my colleagues would always turn the heating up rather than put a jumper on'. Employee responses to these statements accounted for 30 per cent of the final ranking.

For the Green IT initiative to be successful (as with any organisational change programme), sponsorship and support must come from the top. It is therefore not surprising to discover that the six organisations ranked highest by their staff overall also achieved the best six 'my boss' Green scores.

In addition to the social and reputation benefits of the organisation, there are also benefits to be realised from a marketing perspective. An organisation's brand serves to create associations and expectations around it. Therefore most organisations are now recognising that taking the environment seriously and committing to tackling environmental issues has the potential to impact brand value either positively or negatively and are now taking appropriate action. Those organisations which can align their brand to internationally recognised environmental standards and certificates can only gain from such associations. Therefore, adopting initiatives such as Green IT can only serve to enhance the organisation's reputation and brand.

The economic value and cost benefits of Green IT can be realised in many ways. This includes, but is not restricted to, reduced costs provided by initiatives such as teleconferencing and voiceconferencing. These technologies reduce travel costs (in some cases by 20–30 per cent) and improve individual productivity and well-being. An organisation can derive substantial financial benefits from not having its most valuable resource (people) travelling unnecessarily, at considerable financial cost to the business.

Other more complex Green IT solutions (which are covered in more depth later in the book), such as virtualisation, cloud computing, grid computing and thin-client architectures, all help to reduce IT costs and to optimise resources.

8   MAKING GREEN IT HAPPEN

INFORMATION GATHERING AND SETTING TARGETS

There is a well-known quote by Lewis Carroll that states: 'If you don't know where you are going, any road will get you there'.[14] This is certainly true when it comes to achieving the organisation's Green IT ambitions. If the organisation does not have stated targets, and have them documented, well communicated and measured, there is a very high probability that they will never be achieved. Subsequently, without a detailed and comprehensive set of data providing evidence of the current situation, setting targets becomes difficult and in some cases impossible. Therefore the importance of having quantifiable, measurable, well-communicated targets and milestones, positioned against well-defined and accurate baselines, cannot be stressed enough.

As well as defining the current environmental baseline and related targets, it is also vital to gather detailed and accurate information on the cost of current IT service provision, the cost of investment in Green IT and the expected ROI for Green IT. This is especially important for commercially focused organisations as the ROI of a typical investment in Green solutions and technology can be anything between three and ten years, depending on the type of investment being made. However, the focus need not be on financial justification alone. As previously stated when discussing the TBL, the organisation will need to collate evidence on the environmental and political damage being caused by current organisational practices and to highlight the danger of not meeting legislative commitments.

For organisations in the UK, the arrival of the CRC mandates an unprecedented change in working practices, yet many organisations are either unclear or completely unaware of the impact it may have. The CRC is just one of many ways in which the UK government intends to tackle its commitment to reducing carbon emissions, and is a direct result of the passing of the Climate Change Act.[15] The Act was passed in November 2008, and it sets a target for the UK to reduce carbon emissions to 80 per cent below 1990 levels by 2050. It also sets an interim target of a 34 per cent reduction by 2020 (with the potential to increase this to 42 per cent given an international agreement) and establishes the concept of carbon budgets. The Climate Change Act creates a new approach to managing and responding to climate change in the UK by setting ambitious, legally binding targets and by creating powers to help meet those targets.

MAKING GREEN IT HAPPEN

Any organisation that qualifies for the CRC but chooses (intentionally or not intentionally) not to participate, to falsify information or obstruct the regulator could find themselves being prosecuted (see Table 8.1). There are also civil as well as criminal penalties that can be brought against the organisation by the regulator. The CRC is regulated by the Environment Agency in England and the Scottish Environment Protection Agency in Scotland (SEPA). In Northern Ireland the regulator is the Northern Ireland Environment Agency (NIEA).

Table 8.1 Penalties and offences in the CRC

Non-compliance	Penalties
Civil penalties	
Failure to register	• immediate fine of £5,000 imposed for failure to register by the deadline;
	• further fine of £500 per working day for each subsequent working day of delay up to a maximum of 80 working days;
	• publication of non-compliance.
Failure to disclose information	• where an organisation with a Half-Hourly Meter (HHM) that does not meet the qualifying threshold fails to make an information disclosure, a one-off fine of £500 per settled HHM for which that organisation is responsible.
Failure to make a complete registration	• where an organisation registers, but fails to do so on behalf of all parts of their organisation, a fine of £500 per settled HHM for which the organisation is responsible but was not included in its registration;
	• publication of non-compliance.
Failure to provide a footprint report	• immediate fine of £5,000 for failure to provide a footprint report by the reporting deadline;
	• further fine of £500 per working day for each subsequent day of delay up to a maximum of 40 working days. Total accumulated daily rate fine is doubled after 40 working days;
	• publication of non-compliance.

(Continued)

Table 8.1 *(Continued)*

Non-compliance	Penalties
Failure to provide an annual report	• immediate fine of £5,000 for failure to provide an annual report by the reporting deadline; • further fine of £500 per working day for each subsequent day of delay up to a maximum of 40 working days. Total accumulated daily rate fine is doubled after 40 working days. Emissions are doubled only with regard to that year's performance commitment requirement (doubled figure will not count towards the participant's rolling average); • publication of non-compliance; • administrator will block the transfer of all allowances out of the participant's registry account until report is received; • bottom ranking on the league table.
Incorrect reporting	• fine of £40 for each tCO_2 of emissions incorrectly reported – to be applied wherever there is a margin of error greater than 5 per cent; • publication of non-compliance.
Incorrect information	• where an organisation fails to provide accurate information in its reports, and where that information does not affect the emissions totals, a fine of £5,000; • further, where that inaccurate information affects the participant's performance in the league table, an additional fine of double the amount of any financial gain achieved from improved performance score; • publication of non-compliance.
Failure to comply with the performance commitment	• fine of $£40/tCO_2$ in respect of each allowance that should have been obtained and surrendered • must obtain and cancel the outstanding balance of allowances;

(Continued)

Table 8.1 *(Continued)*

Non-compliance	Penalties
(surrendering sufficient allowances)	• continued failure to remedy shortfall will result in recycling payment being withheld until the participant complies; • if a participant fails to comply by the end of the compliance year, they will not receive their recycling payment. Outstanding allowances will then be added to their performance commitment requirement for the following year; • publication of non-compliance; • administrator will block the transfer of all allowances out of the participant's registry account until all necessary allowances are surrendered.
Latent failure to comply with the performance commitment	• where the non-compliant organisation is still a participant, the shortfall of allowances is added to their current compliance year's performance commitment total; • publication of non-compliance; • where the non-compliant organisation is no longer a participant, a fine equal to the value of the shortfall is determined with reference to the price of allowances in the most recent government sale or auction.
Failure to keep adequate records	• fine of £40 per tCO_2 of the participant's CRC emissions in the most recent compliance year; • publication of non-compliance;
Failure of franchisee to provide information to a franchisor	• where a franchisee has failed to provide information to a franchisor which has prevented the franchisor complying with its obligations under the Order, the administrators may impose an enforcement notice on the franchisee.

(Continued)

Table 8.1 *(Continued)*

Non-compliance	Penalties
Criminal offence	
Falsification Knowingly or recklessly make false or misleading statement	• imprisonment up to three months (up to 12 months in Scotland); • fine up to £50,000.
Falsification of evidence Attempt to deceive or mislead the administrator	
Non-compliance with enforcement Failure to comply with an enforcement notice Intentionally obstruct the administrator Failure to provide assistance, facilities and information or to permit any inspection Failure to appear, or prevent any other person to appear, before the administrator as part of an inspection	• indictment; • imprisonment up to two years; • fine (unlimited).

The CRC is the UK's first mandatory carbon-trading scheme, and was kicked off in January 2010. It covers large, non-energy intensive business and public sector organisations. About 5,000 organisations will come within the scope of the new legislation, ranging from retailers, banks, water companies, hotel chains, universities and local authorities. The scheme is compulsory for large organisations using more than 6,000 MWh per year of half-hourly metered electricity, which translates to approximately £500,000 in electricity bills, and is aimed at encouraging large organisations to reduce their fixed source energy consumption. Companies will have to start buying carbon allowances, provided through carbon auctions, to cover their carbon emissions, and that will involve measuring and recording energy use and calculating their CO_2 emissions. Companies using just under 6,000 MWh will have to prove it by putting together an 'evidence pack', which includes details of the organisation's contact names, electricity and fuel bills, automatic metering records, procedures used to collect data and check correctness, and information about special events, such as changes to energy suppliers and organisational subsidiaries. The revenue generated from carbon auctions will then be redistributed between the scheme's participants. Each company will receive a larger or smaller carbon allowance than they originally paid for, according to their performance in the CRC league table.

The main driver for the CRC is to stimulate energy efficiency in organisations which might otherwise be resistant to implementing energy efficiency measures. A league table may affect brand reputation and therefore has a good chance of grabbing management's attention. The CRC league table will be published each year after companies have reported their annual carbon emissions. Participants will be awarded bonus or penalty payments according to their placement in the league. In the first year the repayment adjustments will be between +10 and –10 per cent. This will widen significantly to +50 and –50 per cent by year five. The league table will be based on three metrics – an absolute metric, a growth metric and an early action metric – to gauge the energy saving achievements of companies.

The absolute metric is a measure of your organisation's relative performance in reducing emissions during each annual reporting year. Your performance is determined against your rolling average emissions during the five years prior to the current year. Where five years' data are not yet available, this is the rolling average of all the years of data available up to that point. The growth metric is designed to take into account organisational growth during the five-year phase. Organisations that grow, but with lower emissions intensity, will perform well in this metric.

The growth metric is calculated as the percentage change in emissions per unit turnover (or revenue expenditure for the public sector) against that organisation's annual average emissions per unit turnover (or revenue expenditure). For the purposes of this metric, organisations for which turnover is not relevant, i.e. the public sector, will be able to use a total expenditure figure from their audited accounts that does not include any capital expenditure and which is consistently applied each year – termed 'revenue expenditure'. However, it must cover all the UK operations of the CRC participant.

There are two components to the early action metric (this metric will be removed after the introductory phase). The first component is a calculation of the percentage of the organisation's emissions from electricity and gas that is covered by voluntarily installed Automatic Meter Reading (AMR) in the

2010/11 reporting year.[16] The second component is calculated from the percentage of the organisation's annually reported CRC emissions covered by the Carbon Trust Standard (as well as recognised equivalents) or at the end of each compliance year of the introductory phase. Participants who still hold a current and valid Energy Efficiency Accreditation Scheme (EEAS) – predecessor to the Carbon Trust Standard certificate – at the end of a relevant reporting year will also receive recognition under this metric.[17][18]

The first carbon auction is due in April 2011, when bidders will be able to buy allowances to cover their 2010 emissions as well as to forecast 2011 emissions. Subsequent auctions will only apply to the year immediately ahead. An allowance will need to be purchased for every tonne of CO_2 emitted in the year. The UK government has indicated that the price of carbon is likely to be set initially at £12 per tonne. Using Defra's average electricity/carbon conversion factor (0.523 kg CO_2/kWh), the threshold qualification of 6,000 MWh of electricity translates into mandatory carbon credit purchases of about £38,000. If the organisation uses more than the qualification threshold level, the cash outgoing will be more.

Given the cash flow implications of the CRC, it is essential for those organisations coming within scope to prepare and budget for compliance. Of course, proper data management systems and accurate carbon assessments will be critical for the organisation when compiling the necessary reports, as there will be penalties for under-reporting emissions. This is another compelling example of how critical it will be to the organisation to have robust IT monitoring and reporting in place, as penalties for under-reporting are expected to rise to £75 per tonne from 2013 onwards.

Naturally, there are ways in which organisations can benefit from participation, and those are by delivering what the scheme is designed to achieve, a reduction in carbon emissions. By acting early, implementing energy efficiency measures and reducing carbon reductions, an organisation has the best chance of capitalising on the potential benefits. Other key benefits include reduced energy bills, earning profit by being placed in the upper part of the CRC league table and acquiring the related prestige and positive media attention. The opposites apply, of course, for organisations which do not do well under the CRC and find themselves lower down the league table.

Although revenue from the carbon auctions will be recycled, participants will notice a significant impact on cash flow at the start of the scheme. The baseline year for revenue recycling in the first phase is to be calculated on six months of data, corresponding to the period October 2009 to March 2010. Sales of allowances will take place each April and participants will have to report on emissions. The money generated from the auctions (less the administration costs of running the scheme) will be recycled to the participants. The gap between auctioning and revenue recycling is six months.

The price of carbon may rise in subsequent years, especially as allowances for emissions are restricted. The best way to counter the uncertainties in the price of carbon is to invest in energy-saving measures and to reduce energy usage.

The organisation will then benefit from reduced energy bills and reduced costs with participation in the scheme. An organisation may subsequently drop out of the CRC if it falls below the threshold level. However, to realise the savings, the organisation will need to ensure that it has the right skills to implement the required energy saving measures, and to invest and trade in emissions allowances.

Implementation of the CRC will affect a wide variety of job functions. Energy usage and carbon management activities are often spread across an organisation, from ICT, facilities management, operations, logistics, regulatory affairs, procurement, finance, treasury, environmental, production, corporate social responsibility, health and safety, sustainability and training. The CRC compliance team is likely to interact across all these functions. It would be a mistake to think that compliance should be delivered solely by the ICT or facilities manager.

Participation in the CRC will require expertise in investment and trade in financial instruments for emissions. In Phase 1 of the CRC (from April 2010 to March 2013) participants in the scheme will need to purchase an allowance for every tonne of carbon emitted in the year. In Phase 2 (from April 2013 onwards), the number of available allowances will be capped by the government and the purchase price will be unknown until a sealed bid auction is complete. Because there will be a cap on emissions, it is likely that a secondary market will emerge through which participants can sell and purchase excess allowances. Engaging in auctions and secondary markets for carbon emissions is likely to be new for many organisations and this will highlight a skills gap that needs to be addressed. Organisations can choose either to do some or all of these things internally or to outsource some of them to third parties, such as carbon traders specialising in handling allowance portfolios for their clients.

Some organisations may establish their own teams of carbon traders. Being in a position to at least assemble the evidence pack internally is likely to be cost effective, especially as the organisation evolves its carbon reduction programmes. At the end of the year the organisation will have to calculate its total emissions and submit an evidence pack containing detailed energy records to the government. Given the strategic implications for the organisation, management should be familiar with the CRC and the implications of it for the organisation. Each April (except in year 2 which requires retrospective purchase of allowances), the organisation will need to set an emissions target in order to purchase allowances to cover the expected carbon emissions for the year ahead. Management will also need to give approval for purchasing allowances; therefore it is well worth getting it right.[19]

In conjunction with collating the evidence needed to justify the organisation's investment in Green IT a set of aims will also need to be identified. These can include but not necessarily be restricted to, as has already been stated, reducing the ICT organisation's carbon footprint, helping to reduce the organisation's overall carbon footprint, encouraging Greener behaviours in staff, suppliers and customers and improving the sustainability of resources in the organisation overall.

Once the go-ahead has been given, and the finances have been secured, planning on how the organisation is going to physically build, test and implement the required initiatives will need to take place. This will include having a thorough and detailed understanding of the relevant polices, standards and legislation such as the CRC that will potentially impact the organisation. (See Table 8.2.)

Table 8.2 Key agreements, legislation and guidance

Name	Type	Comments
The UN Framework Convention on Climate Change 1992	International statement of intent	An outcome from the 1992 Rio Earth Summit. A very general agreement to work towards addressing the adverse effects of climate change.
The Kyoto Protocol 1997	International agreement	A supposedly binding agreement by participating countries to meet emission reduction targets for GHGs, by 2012 – in relation to 1990 emission levels. Some participating countries exempted from reduction targets.
Restriction of Hazardous Substances (RoHS) Act	EU law	Aimed at eradicating the use in new electrical/electronic equipment of hazardous substances such as lead, mercury, cadmium etc.
Waste Electrical and Electronic Equipment (WEEE) Directive	EU law	Aimed at reducing production and use of electrical and electronic equipment. Regulates disposal of equipment and recovery of raw materials.

(Continued)

Table 8.2 *(Continued)*

Name	Type	Comments
European Batteries and Accumulators Regulation 2009	EU law	Regulates manufacture and especially disposal of batteries, e.g. businesses selling batteries are now also obliged to provide recycling arrangements.
UK Climate Change Act 2008	UK law	Legal framework to allow achievement of UK emission-reduction targets. Supported by CRC and Carbon Budget System (implemented in 2009).
BSi Publicly Available Specification (PAS) 2050	Guidance	An assessment, reporting and labelling scheme that highlights embodied GHG emissions in goods and services. Joint initiative between Defra and the Carbon Trust.
EU Code of Conduct on Data Centre Energy Efficiency	Guidance	Aimed at guiding data-centre operators towards greater efficiency – particularly in respect of reducing energy consumption.
Electronic Product Environmental Assessment Tool (EPEAT)	Assessment tool	Product assessment scheme allowing consumers to check the environmental impact of electronic products. Products are categorised into bronze, silver or gold levels.
ISO 14000, 14001 etc.	Standard	A family of general environmental management standards (not specifically IT-related).

UNDERSTANDING THE ORGANISATION'S CARBON FOOTPRINT

What is also critical to the success of the Green IT plan is for organisations to have a thorough understanding of the carbon footprint associated with ICT. If this is not known, then it will be difficult for organisations to provide accurate reporting on improvements without a baseline to measure against. Of course, for organisations whose size and electricity consumption means that they are subject to the requirements of the CRC, this is a legal requirement; for other organisations, it is key to actively demonstrating their Green credentials.

Unfortunately, identifying a carbon footprint calculator that suits the organisation's particular needs is not a simple exercise. For numerous organisations, the use of more than one calculation tool may be required as there are countless different types, measuring various different aspects. Several organisations have adopted the GHG Protocol – perhaps the most widely used international accounting tool, which has been built to allow government and organisation leaders to understand, quantify and manage GHG emissions. The first edition of *The Greenhouse Gas Protocol: A Corporate Accounting and Reporting Standard* (Corporate Standard) was published in 2001. Since then, the GHG Protocol has built upon the Corporate Standard by developing a suite of calculation tools to assist companies in calculating their GHG emissions and additional guidance documents such as the GHG Protocol for Project Accounting.

To complement the standard and guidance, a number of cross-sector and sector-specific calculation tools are also available, including a guide for small, office-based organisations. These tools provide step-by-step guidance and electronic worksheets to help organisations calculate GHG emissions from specific sources or industries. These tools are consistent with those proposed by the Intercontinental Panel on Climate Change (IPCC) for compilation of emissions at a national level. Thanks to an intensive review by many companies, organisations and individual experts, the tools are believed to represent current 'best practice'. However, none of them is ICT-specific.

Regardless of the type of calculation tool an organisation decides to use, a core set of principles must underpin all GHG calculations. The Corporate Standard describes these as:

- **Relevance** Ensure that the GHG inventory appropriately reflects the GHG emissions of the company and serves the decision-making needs of users – both internal and external to the company.

- **Completeness** Account for and report on all GHG emission sources and activities within the chosen inventory boundary. Disclose and justify any specific exclusions.

- **Consistency** Use consistent methodologies to allow for meaningful comparisons of emissions over time. Transparently document any changes to the data, inventory boundary, methods or any other relevant factors in the time series.

- **Transparency** Address all relevant issues in a factual and coherent manner, based on a clear audit trail. Disclose any relevant assumptions and make appropriate references to the accounting and calculation methodologies and data sources used.

- **Accuracy** Ensure that the quantification of GHG emissions is systematically neither over nor under actual emissions, as far as can be judged, and that uncertainties are reduced as far as is practicable. Achieve sufficient accuracy to enable users to make decisions with reasonable assurance as to the integrity of the reported information.

GHG accounting and reporting practices are continually evolving and are new to many businesses; however, the principles listed above are derived in part from generally accepted financial accounting and reporting principles. They also reflect the outcome of a collaborative process involving stakeholders from a wide range of technical, environmental and accounting disciplines.

The Corporate Standard also provides guidance on how to set your GHG reduction target, and whether that target should be an absolute or intensity target. An absolute target is usually expressed in terms of a reduction over time in a specified quantity of GHG emissions. For example, 'reduce the GHG emissions associated with the UK data centre by 5 per cent per year from 2008 to 2012'. An intensity target is usually expressed as a reduction in the ratio of GHG emissions relative to another business metric. For example, 'reduce the GHG emissions associated with the UK data centre by 10 per cent per terabyte of data stored'. The comparative metric should be carefully selected. It can be the output of the organisation or service provider or some other metric such as sales, revenues or office space. To facilitate transparency, companies using an intensity target should also report the absolute emissions from sources covered by the target. Table 8.3 summarises the advantages and disadvantages of each type of target.

Table 8.3 Comparing absolute and intensity targets as defined in the GHG Protocol Corporate Accounting and Reporting Standard

Type of target	Advantages	Disadvantages
Absolute	Designed to achieve a reduction in a specified quantity of GHGs emitted to the atmosphere.	Target base year recalculations for significant structural changes to the organisation add complexity to tracking progress over time.

(Continued)

Table 8.3 *(Continued)*

Type of target	Advantages	Disadvantages
	Environmentally robust as it entails a commitment to reduce GHGs by a specified amount.	Does not allow comparisons of GHG intensity/efficiency.
	Transparently addresses potential stakeholder concerns about the need to manage absolute emissions.	Recognises a company for reducing GHGs by decreasing production or output.
		May be difficult to achieve if the company grows unexpectedly and growth is linked to GHG emissions.
Intensity	Reflects GHG performance improvements independent of organic growth or decline.	No guarantee that GHG emissions to the atmosphere will be reduced – absolute emissions may rise even if intensity goes down and output increases.
	Target base year recalculations for structural changes are usually not required.	Companies with diverse operations may find it difficult to define a single common business metric.
	May increase the comparability of GHG performance amongst companies.	If a monetary variable is used for the business metric, such as euro of revenue or sales, it must be recalculated for changes in product prices and product mix, as well as inflation, adding complexity to the tracking process.

Once the target type has been agreed, the next step is to decide on the target boundaries. The factors which organisations will need to consider include agreeing which GHGs are to be reported on and which direct and indirect emissions are going to be included. If the organisation is spread across different geographic sites, will the sites be included all together or will they be treated as different business units or types?

Further to agreeing the target boundaries, organisations will need to decide on a target base year to measure against. The options are to use either a fixed target base year or a rolling target base year. A typical fixed target base year measure will be well defined; for example, 'the organisation will aim to reduce its GHG emissions by 5 per cent in 2010 compared to emissions from 2009'. For a rolling-based target, however, the target may take the form of 'for the next five years starting in 2010 the organisation will aim to reduce its GHG emissions by 5 per cent compared to the previous year'. Organisations may consider using a rolling target base year if to obtain and maintain reliable and verifiable data for a fixed target base year is likely to be challenging (for example, due to frequent acquisitions). With a rolling target base year, the base year rolls forward at regular time intervals, usually one year, so that emissions are always compared against the previous year.

Once the target base year has been agreed, the target completion date also needs to be agreed. A long-term target completion date will enable the organisation to plan ahead and to plan for large capital investments. However, it may also lead to a delay in retiring inefficient legacy equipment. Also, there are risks associated with not knowing or being able to predict the long-term future of the organisation that in turn will make the planning of GHG emission reduction initiatives more difficult. In association with agreeing the target completion date, the organisation will also need to define the length of the commitment period. This is the period of time during which emissions performance is actually going to be measured and ends with the target completion date. For example, the commitment period of the Kyoto Protocol is five years (2008–2012).

The next step for organisations to undertake is to decide on the use of carbon offsets or carbon credits to assist in meeting their carbon emission reduction targets. Carbon offsetting allows an organisation to compensate for its unavoidable emissions by helping to fund projects that deliver an equivalent CO_2 saving elsewhere. Once you have decided to offset, you can choose to purchase an approved offset that meets a specific quality criteria. A source of approved offsets is the UK government's carbon quality assurance scheme. The scheme ensures that the offset provider will sell good quality carbon credits that comply with the Kyoto Protocol and have been verified by the United Nations (UN) or the European Union's (EU's) emission trading scheme. The use of offsets may be appropriate when the cost of making internal reductions is prohibitive and previously planned opportunities become difficult to deliver, perhaps due to major organisational change. However, the organisation should make every effort to directly reduce its emissions, before resorting to offsets.

With the introduction of offset schemes, organisations need to have a policy to deal with the possibility of double counting of reductions. This can occur when a GHG offset is counted towards the target by both the seller and purchaser of the offset. For example, company A undertakes an internal reduction project that reduces GHGs that are then included in their own target. Company A then also sells the reduction to company B to use as an offset towards its target, whilst still counting it toward its own target. This in turn means that the reductions are counted by two different organisations against targets that cover different emissions sources. Recognised trading programmes, such as the UK Government Quality Assurance Scheme mentioned above, address this by using registries that allocate a serial number to all traded offsets or credits and by ensuring that the serial numbers are retired once they are used. In the absence of registries this could be addressed by a contract between seller and buyer.

The penultimate step when setting up a GHG target is to decide on the target level. Key points at this stage include understanding what is business-as-usual, how far to go beyond that and being clear on how all the above steps influence the decision. The decision on setting the target level should be influenced by all the previous steps. A useful means of helping an organisation decide on the target level is to use a simple GHG balance sheet to track performance towards the target. The balance sheet should show for each commitment period the emissions cap for the organisation, both regulatory and voluntary, the number of allowances purchased and the total balance. See Table 8.4.

The final step once the target has been set is to track performance against the target. This last step is essential in order to check compliance, to produce regular reporting and to provide a means of assessment and comparison to feed into an ongoing improvement plan. It is important to link the target to the annual GHG inventory process and to make regular checks of emissions in relation to the target. Some companies use interim targets for this purpose (a target using a rolling target base year automatically includes interim targets every year). Organisations should include the following information when setting and reporting progress in relation to a target:

- description of the target:

  o provide an outline of the target boundaries chosen;

  o specify target type, target base year, target completion date and length of commitment period;

  o specify whether offsets can be used to meet the target; if yes, specify the type and amount;

  o describe the target double-counting policy;

  o specify the target level;

- information on emissions and performance in relation to the target:

  o report emissions from sources inside the target boundary separately from any GHG trades;

  o if using an intensity target, report absolute emissions from within the target boundary separately, both from any GHG trades and from the business metric;

o report GHG trades that are relevant to compliance with the target
(including how many offsets were used to meet the target);

o report any internal project reductions sold or transferred to another
organisation for use as an offset;

o report the overall performance in relation to the target.

Table 8.4 GHG balance sheet (all values measured in tonnes CO_2 e/year)

GHG target description	GHG emissions	Actual figure + or –	Comments
Voluntary cap (direct emissions)	1,110		
Regulatory cap (direct emissions)	3,300		
TOTAL CAP	4,410	7,720	
Sum of remaining emissions	3,310		
Clean Development Mechanism (CDM) credits purchased (+)	1,200+		
Regulatory allowances purchased (+)	4,000+		
Sum of allowances and credits	5,200		
Sum of allowances and credits minus remaining emissions	1,890+		

9   THE ICT LIFE CYCLE

With organisations, households and individuals now heavily reliant on IT, it is crucial that we understand not only the environmental impact of the energy consumption of infrastructure, but also the environmental impact of its manufacture, transport, usage and disposal. It is therefore vital that we consider not only the GHG emissions associated with energy **consumption**, but also the **embodied** emissions.

Embodied energy is the energy and associated carbon emissions relating to the mining of minerals and the production, transport, manufacturing and disposal of components and devices. Consumed energy, however, is the energy and therefore carbon emissions associated with the use of a device over its operating life. See Figure 9.1.[20]

Figure 9.1 Key stages in the ICT life cycle

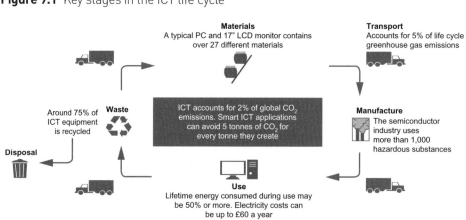

All IT hardware relies on energy usage both in its production as well as in its operation. This energy usage coupled with other elements of production, for example, the mining of precious metals, has a detrimental effect on the environment. Many different minerals, precious metals, chemicals and locations are likely to be used in the production life cycle of a typical electronic notebook.

Whilst it is recognised that the manufacturing of electronic and IT equipment such as PCs and notebooks is a cause of potential environmental damage, we also need to consider the other elements of the life cycle such as transport and disposal. A Gartner Group report in 2008 stated that the number of PCs in use across the globe has surpassed the one billion figure. New and emerging markets such as Brazil, China and India are expected to continue to drive the market's rapid growth and the number of PCs in use on a global scale could easily reach two billion by 2014. Gartner surmises that the prediction is accurate due to continually falling prices and the perception that computers are indispensable for economic advancement.[21]

The rampant growth in demand is further fuelled by the continued advancements in technology such as wireless connectivity and the continually falling price of technology to the average consumer. The demand for technology is further exacerbated by the perception, globally, that nothing can be achieved without the use of technology.

Inextricably linked to this huge demand for technology is not only its manufacture but also its transport. As we have already surmised, a large proportion of technology manufacturing takes place outside the country of its intended use. Therefore, when calculating the embodied emissions of a PC, notebook, printer or any other piece of infrastructure, the emissions directly attributed to transport have to be considered, including the transport from the supplier to its final destination.

Figure 9.2 Ecological damage caused by ICT landfill

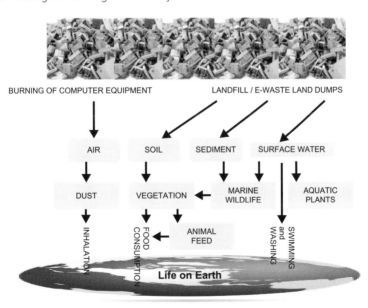

Finally, when considering the combined GHG emissions associated with technology, it is vital that we also include the disposal of said equipment. Globally, millions of PCs, notebooks, printers, servers, mobile devices and associated peripherals will be replaced every year, leading to the unwanted equipment taking various different journeys. Gartner expects more than 180 million computers will be replaced in 2010, with some sold to second owners through various channels, some broken up and recycled, but many will simply be dumped directly into landfill.

The last of these options potentially causes the biggest environmental impact. Electronic ICT equipment contains large numbers of both toxic and non-toxic materials. Once the equipment has been condemned to landfill the leaking of poisonous and toxic materials such as lead and asbestos contaminates the surrounding areas, and over time leads to pollution and contamination of both soil and water, and eventually the whole eco-system. Ultimately this culminates in the poisonous and toxic substances finding their way into both the animal and human food chains (Figure 9.2).

10 HOW TO LESSEN THE IMPACT OF EMBODIED EMISSIONS

Considering the potential environmental impact, it is imperative that both organisations and individuals implement solutions to reduce the embodied GHG emissions of ICT equipment. These initiatives include (but are not restricted to) the following:

FIT FOR PURPOSE E-WASTE DISPOSAL

There is a real and potential danger for organisations to find themselves on the wrong side of the law as well as causing environmental damage through negligent e-waste disposal. Under legislation such as the Waste Electrical and Electronic Equipment (WEEE) Directive, if you are a producer, re-seller or importer of electronic equipment, you will have legal obligations with regard to the recycling, reuse and disposal of electronic equipment or components. If you are a business user with electronic waste, the UK Environment Agency offers the following useful and relevant advice.[22]

New regulations have been introduced to tackle the growing amount of WEEE. It is one of the fastest growing waste streams in Europe, and, in the UK alone, we throw away around 2 million tonnes of WEEE every year, much of which ends up in landfill. The regulations aim to ensure that more WEEE is separately collected for treatment and recovery, and less goes to landfill.

The regulations apply to all companies who import, manufacture or re-brand electrical equipment in the UK; these companies are known as 'producers'. They also affect everyone who uses, sells, treats or disposes of WEEE. They affect the way WEEE is disposed of by setting treatment standards and recycling targets and, importantly, by making producers, rather than end-users, pay for its treatment and recycling in most cases.

Implications for business users
Shifting the burden of payment for the treatment, recycling and disposal of WEEE from end-users to producers will have a significant impact on purchasing and disposal arrangements.

Responsibility for business WEEE
If you bought equipment before 13 August 2005, and are replacing it with new equipment fulfilling the same function, then the producer of the new equipment

is responsible for the collection, treatment and recycling of the old equipment, regardless of whether they were the original manufacturer. If you bought the equipment before 13 August 2005 and do not replace it, then you are responsible for financing and arranging treatment in accordance with the WEEE Regulations and existing waste management legislation, including the Duty of Care and the Hazardous Waste Regulations.

If you bought electrical equipment after 13 August 2005, then the producer of that equipment is responsible for its collection, treatment and recycling when you dispose of it. If you lease or rent equipment, the producer is usually responsible for its disposal. The regulations allow producers and business users to agree 'alternative arrangements', whereby the business user agrees to take on some or all of the future costs of the end-of-life treatment of the equipment he or she buys. This is a commercial decision that you will need to make and is likely to form part of the normal negotiating processes for supply contracts in the future.

Collection arrangements
WEEE from business users may be collected by the obligated producer or the compliance scheme working on its behalf. WEEE may either be collected directly from your premises or you may be asked to take the WEEE to a local collection facility (which should be easily accessible to you).

What you need to do
Think about the environmental consequences before deciding to replace equipment. Do you really need to buy a new product? Your equipment could be upgraded or you could buy a refurbished product instead. If you have working equipment that you no longer need, think about passing it on to others instead of throwing it away. Remember to take account of the WEEE Regulations when entering into commercial negotiations and procurement decisions concerning Electrical and Electronic Equipment (EEE).

If you are buying electrical products from a distributor or other intermediary, make sure that you obtain the producer registration number for the equipment being supplied so that you know who to contact to arrange disposal at the end of its life. Suppliers of EEE should always be able to provide this on behalf of producers.

If the producer is responsible for disposal, establish whether it is the producer or the producer's compliance scheme who will be arranging for disposal, and whether collection will be from your premises or a local collection facility. Organisations will also need to be aware of their responsibilities under other waste management legislation, such as the Duty of Care and the Specialist Waste Regulations.[23]

REDUCE THE RELIANCE ON RECYCLING IN DEVELOPING AND THIRD WORLD COUNTRIES

As both organisations and individuals in what may be referred to as developed countries adapt to modern ways of working in the 21st century; there are many

examples of 'western' consumerism and growth leading to exploitation in
Third World or developing countries. An example of this is the Chinese town
of Guiyu.

Guiyu is in the Guangdong province of China and is now practically synonymous
with WEEE dumping from the West. It is described as one of the most toxic
places on Earth, where the levels of lead recorded in its citizens' blood far exceeds
the levels found even in neighbouring towns. The tragedy of this is that the
majority of the electronic waste found here is present as a result of loopholes in
EU and US legislation. For example, under EU law, used electronic equipment
can be exported if it is still working; however, when organisations such as
Greenpeace have investigated further, it is obvious that significant amounts of
WEEE are being exported illegally and end up in waste dumps such as Guiyu
and in West African countries such as Ghana and The Ivory Coast. Guiyu was
first brought to the attention of the world's media in December 2001 by the Basel
Action Network in their report and documentary film entitled Exporting Harm:
the High Tech Trashing of Asia.[24]

Since then, Guiyu has attracted attention from many different organisations
including Greenpeace and the UN Environmental Programme. Working practices
discovered in Guiyu by the Basel Action Network included:

Toner sweeping
Workers without any specialist or protective clothing opened discarded toner
cartridges with screwdrivers and then used paint brushes and their bare hands
to wipe the toner into a bucket. The process creates constant clouds of toner that
billowed around the workers and was routinely inhaled. In the course of the
working day, the worker's skin and clothing was blackened.

Open burning
This is the process of burning wires in open fires to retrieve the copper that
exists inside them. The emissions and ashes from such burning will contain
high levels of both brominates and chlorinated dioxins and furans – two of the
most deadly Persistent Organic Pollutants (POPs). It is also highly likely that
cancer-causing Polycyclic Aromatic Hydrocarbons (PAHs) are also present in the
emissions and ash.

Cathode Ray Tube (CRT) cracking and dumping
Copper-laden yokes from the ends of the CRTs are broken off, with the CRT
itself being cracked and discarded in the process. The yokes are then sold to
copper-recovery operations.

Circuit board recycling
It is likely that the most environmentally destructive recycling overall involves
the recovery of the various components and materials found on electronic circuit
boards. The general approach to recycling a circuit board involves first a
de-soldering process. Many hundreds of workers, usually women and girls, are
active each day in this endeavour. They place the circuit boards on shallow
wok-like grills that are heated underneath by a can filled with ignited coal. In the
wok grill is a pool of molten lead-tin solder. The circuit boards are placed in the

pooled solder and heated until the chips are removable. These are then plucked out with pliers and placed quickly in buckets. Solder is also collected by slapping the boards hard against something such as a rock where the solder collects and is later melted off and sold. Whilst fans are sometimes used to blow the toxic lead-tin solder fumes away, the exposure on a daily basis is likely to be very damaging. The loosened chips are then sorted between those valuable for resale and those to be sent to the acid chemical strippers for gold recovery. Often, the pins on chips will be straightened and later dipped in fresh solder to make them look new for use in the computer re-fabrication business.

After the de-soldering process, the stripped circuit boards go to another less skilled labourer who then removes capacitors and other less valuable components for separation with wire-clippers. After most of the board is picked over, it then goes to large-scale burning or acid recovery where the last remaining metals are recovered. This final burning process is bound to emit substantial quantities of harmful heavy metals, dioxins, beryllium and PAHs.

Acid stripping of chips

Much of the work to remove chips from circuit boards is done for the ultimate purpose of removing precious metals. This is most often done by a very primitive process using acid baths. The baths are thought to contain aqua regia (a mixture of 25 per cent pure nitric acid and 75 per cent pure hydrochloric acid). The aqua regia is first heated over small fires and then poured into plastic tubs full of computer chips. These in turn are routinely swirled and agitated to dissolve the tiny amounts of gold found inside. After many hours of this, a chemical is then added which precipitates the gold, making it settle to the bottom of the tub. This is recovered as a mud, dried, and then finally melted to a tiny bead of pure, shiny gold. The process results in huge clouds of steamy acid gases being emitted. The process also results in the routine dumping of aqua regia process sludge that blackens the river banks with the resinous material making up computer chips. A quick test using pH paper on the saturated ground surrounding the tubs measured a pH level of 0, the strongest level of acidity.

The men working on this process day and night are protected only by rubber boots and gloves. They have nothing to protect them from inhaling and enduring the acrid and often toxic fumes. The aqua regia process is known to emit toxic chlorine and sulphur dioxide gases.

Plastic chipping and melting

The plastic parts of e-waste and, in particular, the housings of computers, monitors and plastic keyboard parts are all sent to one of the Guiyu villages that is involved with processing plastics. Much time is spent there chipping plastics into small particles and then separating the various colours of plastics so that a clean coloured re-melt is possible. Then the chips are bagged and sent to melting and extruding operations. Often children are employed for this tedious job and the melting of the computer plastics is done in rooms with little ventilation and with no respiratory protection. It is not even known if respiratory protection equipment would be capable of filtering out the dangerous hydrocarbons, including the dioxins and furans, that are likely to be produced when melting brominates flame-retardant-impregnated plastic or PVC plastic.

Despite the attempt to recycle much of the plastic from the e-waste stream, a large percentage is deemed unrecyclable due to impurities or the difficulty in separating it, or matching the colours. The result of this is that many, many tons of plastic e-waste are dumped throughout the landscape and most often near waterways.

Materials dumped

A tremendous amount of imported e-waste material and process residues are not recycled but simply dumped in open fields, along riverbanks, ponds, wetlands, in rivers and in irrigation ditches. These materials include leaded CRT glass, burned or acid-reduced circuit boards, mixed, dirty plastics, toner cartridges and considerable material apparently too difficult to separate. Also dumped are residues from recycling operations including ashes from numerous open burning operations, and spent acid baths and sludges. It is this indiscriminate dumping which has no doubt led to the severe contamination of the drinking water supply of Guiyu.

PURCHASING AND SOURCING OF ENVIRONMENTALLY FRIENDLY EQUIPMENT

As stricter regulations and governance is introduced in relation to the manufacture, use and disposal of ICT equipment, more guidance and advice is being sought before the purchase and procurement of ICT equipment by organisations and governments globally. One of the most recent and popular initiatives that is helping organisations and individuals to make the right choice is the EPEAT®.

The development of EPEAT was prompted by a growing demand for an easy-to-use evaluation tool that enables the comparison and selection of electronic products based on environmental performance attributes. IT purchasers needed a simple way to assess products' environmental impacts, and electronics manu-facturers in turn wanted consistent guidance to ensure that their Green design efforts met with success in the marketplace.

EPEAT meets both constituencies' needs with a user-friendly system designed and guided by all stakeholders that is accessible to purchasers and manufactur-ers of any size. As a result, EPEAT has revolutionised the electronic product sector, with significant manufacturer and purchaser participation and an extensive registry of hundreds of electronic products that meet the system's demanding criteria.

EPEAT is a multi-dimensional environmental performance standard for electronic products that currently (2010) covers desktops, notebooks, worksta-tions, thin client devices and displays. Standards development is underway for Imaging Equipment and Televisions, with Servers and Mobile Devices standards development planned for 2011. Widely known and broadly used, EPEAT helps manufacturers to gain market recognition for their environmental efforts. Increasingly, EPEAT is a condition of market access.

EPEAT was developed through a three-year stakeholder consensus process underwritten by the US EPA that included environmental advocates, manufacturers, public and private purchasers, electronics recyclers and technology researchers. The EPEAT standard was formally adopted through an American National Standards Institute (ANSI) accredited process in 2006 as Standard 1680 of the Institute of Electrical and Electronic Engineers (IEEE).

The original participants in the standard development process were primarily USA and Canada-based; however, participation has widened since EPEAT's launch to include representatives of government and private enterprise from the EU, Asia and Latin American regions.

From its launch in July 2006 with three participating manufacturers and a roster of 60 registered products, the EPEAT system has grown to include over 40 manufacturer participants registering more than 1,200 environmentally preferable ICT products in more than 40 countries worldwide. In August 2009, the EPEAT system rolled out to 38 countries in addition to the USA and Canada, to meet international purchasers' demands for EPEAT registrations under local product numbers and with programme support in geographies outside the US market. For a complete list of all participating countries, refer to the country list at http://www.epeat.net/International.aspx.

The EPEAT system evaluates electronic products according to three tiers of environmental performance, bronze, silver and gold, based on a total of 51 environmental criteria. Products that meet 23 required criteria may register in EPEAT; they then qualify for silver or gold ratings by meeting increasing percentages of 28 optional criteria. Stakeholders laid out a system of eight categories of environmental attributes or impacts, with between four and eight criteria in each one. Some criteria overlap with other standards like ENERGY STAR or regulatory regimes like the Restriction of Hazardous Substances (RoHSs); others identify new, cutting-edge product attributes that have not yet been addressed in any other standard, and others reward incremental improvement over a standard baseline. Key benchmarks include elimination of toxic materials, design for recycling, extended product longevity, increased energy efficiency and availability of take-back and recycling services. The full criteria list is available at http://www.epeat.net/Criteria.aspx.

EPEAT is a purchasing requirement for all US Federal Agencies, and is integrated into hundreds of government, education, healthcare and enterprise IT contracts worldwide. In addition, EPEAT is becoming an integral part of the IT channel, with leading distributors and resellers joining EPEAT Partner programs to support their customers' Green purchasing initiatives. Consumer and small business outlets such as Amazon.com and Best Buy for Business provide EPEAT identification for their customers' use, identifying 'Greener' products. Global usage of EPEAT is increasing, and expansion to additional geographies and product sets is in the pipeline.

Perhaps the most intriguing aspect of EPEAT is its demonstration that voluntary compliance and variable declarations can effectively move the market towards more sustainable products. As opposed to conventional single-tier certification systems, EPEAT does not establish a one-size-fits-all route to certification. Instead, by offering multiple levels of certification and flexibility in meeting the optional criteria needed to rank higher, EPEAT has harnessed the competitive forces of the market to drive product and service redesign. This aspect of the system has made it particularly well suited to the fast-moving electronic product development arena – since optional criteria may be included in the standard that are as yet unmet, or met by only a very few products.

When the program was launched in July 2006, no products were rated EPEAT gold. It was not until a year later (after significant work on the part of subscribing manufacturers) that the first gold desktop and notebook were registered by HP and Dell respectively. Since then, the numbers of silver and gold products have risen quickly, with over 400 products now qualifying for EPEAT gold status. Many of the optional criteria show very high conformance rates (i.e. inclusion in 75–85 per cent of product declarations) – demonstrating that mandatory criteria are not the only way to encourage design and service improvement.

Conformance with the EPEAT criteria is declared by the subscriber companies themselves, through a web-based registration tool (Figure 10.1), with regular rounds of verification investigation by the EPEAT organisation ensuring quality control over those declarations. Subscribing manufacturers must be ready at any time to back up a specific declaration – for one or several criteria, or an entire product record – as verification investigations are launched without forewarning, and the results of all investigations are published. This has posed a unique challenge to subscribers, in that they must have internal verification and documentation processes for all criteria established ahead of their actual declaration. (See EPEAT Conformity Assessment Protocols at http://www.epeat.net/VerificationProtocols.aspx for the specific evidence which subscribing manufacturers may be required to provide to support specific criteria declarations.) In an industry with such a complex supply chain, these processes can require a great deal of work and surveillance.

EPEAT's unconventional approach – product declaration by the manufacturer, followed by registry surveillance and ongoing verification investigation – was decided upon by the stakeholders during development of the IEEE 1680 standard.

The group very carefully considered the most effective way to maintain the credibility of the registry based on the unique characteristics of these high-tech products:

- very rapid technology development;
- very short time to market;
- very complex and continually morphing global supply chains;
- very high variability in the configurations of individual products (components from totally different supply chain manufacturers may be found inside the 'same product' over time).

Anecdotally, manufacturers indicate that electronic and computer products experience as high as a 70 per cent rate of changes in components, sourcing and

Figure 10.1 The EPEAT web registration tool

	EPEAT BRONZE	EPEAT SILVER	EPEAT GOLD	Totals
Desktops	0	25	37	62
Displays	0	33	24	57
Integrated Desktop Computers	0	5	9	14
Notebooks	0	118	57	175
Thin Clients	0	13	0	13
Workstation Desktops	0	3	7	10
Workstation Notebooks	0	1	0	1
Total:	**0**	**198**	**134**	**332**

other elements from the original product launch through the commercial life of a given model. Given this rate of change, a pre-certification based on a one-time investigation before a product is on the market is fundamentally inadequate to assess IT equipment as it will be delivered to the purchaser. Stakeholders recognised that ongoing and randomly timed surveillance would be the best way to identify potential problems.

Therefore, in accordance with the IEEE 1680 standard, EPEAT has developed rigorous and transparent post-declaration verification procedures based on unannounced and very in-depth investigations, and on public exposure in the case of non-conformances. The system is designed to make non-conformance publicly embarrassing, and to maintain the constant likelihood of investigation at any time.

Current EPEAT end users include the following US and Canadian states and Provinces: California, Maine, Massachusetts, Minnesota, New York, Oregon, Pennsylvania, Washington, Wisconsin, Nova Scotia, Ontario and Quebec. US and Canadian cities include: LA County, Phoenix, Portland, San Francisco, San Jose and Vancouver.

In the UK, Leeds City Council is a recognised EPEAT user and from an organisational perspective, DeLoitte, Fairmount Hotels, HDR, Kaiser Perma-nente, KPMG and Marriott International are amongst the adopters of EPEAT.

Globally, national government users include the Canadian Federal Government, and national government agencies in Australia, Lithuania, New Zealand, Poland, Singapore and the UK.

Environmental benefits of EPEAT purchases

To enable purchasers to measure the benefits of their EPEAT purchasing in comparison with purchase of conventional products, the US Environment Protection Agency (EPA) supported the development of a life cycle environmental benefits calculator by the University of Tennessee Centre for Clean Products and Clean Technologies. The calculator assesses environmental benefits from electronic product purchases based on specific EPEAT criteria and tiers. By entering information provided by EPEAT's subscribing manufacturers on unit sales of registered products, it is possible to estimate the environmental benefits of overall EPEAT purchasing year by year. It is worth remembering that EPEAT is not the sole motivator of the environmental benefits. In addition to some unique criteria of its own, the EPEAT system brings together into one unified tool such critically important criteria as ENERGY STAR and Restriction of Hazardous Substances compliance, and attributes required under other major environmental evaluation programmes.

Although EPEAT was only developed and made available in 2006, by 2007 sales of EPEAT-registered products worldwide already totalled more than 109 million individual units. This is a demonstration of the rapid growth of EPEAT products' market share. In the same period, EPEAT-registered desktop and laptop sales constituted more than 22 per cent of total worldwide units shipped. Despite the economic downturn and generally flat computer sales in 2008, more than 44 million EPEAT-registered products were sold in the USA, an increase over 2007 of more than a million units, with a very significant (57 per cent) increase in sales in the notebook category paired with declines in both desktops and monitor and display units.

The following findings are proof of the life cycle environmental benefit of EPEAT sales, compared to the purchase of conventional products.[25] EPEAT was able to:

- reduce the use of toxic materials, including mercury, by 1,021 metric tons, equivalent to the weight of 510,949 house bricks;
- eliminate use of enough mercury to fill 149,685 household medical thermometers;
- avoid the disposal of 43,000 metric tons of hazardous waste, equivalent to the weight of almost 22 million house bricks;
- eliminate 14,353 metric tons of solid waste, equivalent to what 7,202 US households generate in a year.

In addition, due to EPEAT's requirement that registered products meet ENERGY STAR's energy efficiency specifications, these products will consume less energy throughout their useful life, resulting in:

- savings of over 8.39 billion kWh of electricity — enough to power over 700,000 US homes for a year;
- reduction in the use of 14.8 million metric tons of primary materials, equivalent to the weight of more than 114 million refrigerators;

- avoidance of 34.2 million metric tons of air emissions (including GHG emissions) and over 71,000 metric tons of water pollutant emissions;
- reduction of over 1.57 million metric tons of GHG emissions — equivalent to taking over one million US passenger cars off the road for a year.

The immense volume of EPEAT registered products sold in 2008, and the very significant environmental and financial benefits resulting from this, confirm the EPEAT system's success as a driver for environmental change in the electronic products market. Credit for these benefits goes to the many purchasers who are demanding EPEAT products, and to the manufacturers who are developing products and services to meet EPEAT's requirements and reduce environmental impact.

Year on year since its conception in 2006, the Green Electronics Council (GEC) anticipates robust continued growth in EPEAT product registrations.[26] As more products are designed to meet the current EPEAT standard, as the current computer standard is updated and as standards covering additional electronic products come on line, these tangible benefits will continue to grow. Finally, EPEAT's expansion from a single registry to one that encompasses 40 countries will enable purchasers worldwide to buy more EPEAT registered products more easily, increasing the EPEAT system's impact over the coming years.

In association with EPEAT, other assessment tools and initiatives such as ENERGY STAR should be considered. ENERGY STAR began life as a joint program of the US Environmental Protection Agency and the US Department of Energy, to promote energy-efficient products and practices both in the home and for businesses and organisations. In 1992 the US EPA introduced ENERGY STAR as a voluntary labelling program designed to identify and promote energy-efficient products to reduce GHG emissions. Computers and monitors were the first labelled products. In 1995, EPA expanded the label to additional office equipment products and residential heating and cooling equipment. In 1996, EPA partnered with the US Department of Energy for particular product categories.

The ENERGY STAR label can potentially appear on all major electrical appliances, office equipment, lighting, home electronics and more. EPA has also extended the label to cover new homes and commercial and industrial buildings.

Through its partnerships with more than 15,000 private and public sector organisations, ENERGY STAR delivers the technical information and tools that organisations and consumers need to choose energy-efficient solutions and best management practices. It is estimated that in the USA, ENERGY STAR successfully delivered energy and cost savings to businesses, organisations and consumers of about $19 billion in 2008 alone. Over the past decade, ENERGY STAR has been a driving force behind the more widespread use of such technological innovations as efficient fluorescent lighting, power management systems for office equipment and low standby energy use, all of which can have a direct or indirect contribution to an organisation meeting its Green IT CRCs.

Although the ENERGY STAR program is very much a US initiative, in 2006 agreement was achieved between the US Government and the European Community (EU) to coordinate energy labelling of office equipment. It is generally accepted that the manufacturing, purchasing and use of office equipment is a huge growth area for both consumers and business organisations. For the purposes of the agreement, office equipment is identified as computers, computer monitors, photocopiers, printers, digital duplicators, faxes, franking machines, multifunction devices and scanners.

As new and exciting innovations in electronic office equipment are realised, consumers and businesses alike are planning upgrades, new purchases and replacements of numerous types of electronic office equipment. Therefore, the Act that led to the coordination of energy-efficiency labelling programmes for office equipment between the US Government and the European Community is seen as a big step forward.

The Act has five general principles:

(i)   A common set of energy-efficiency specifications and a common logo shall be used by the Parties for the purpose of establishing consistent targets for manufacturers, thereby maximising the effect of their individual efforts on the supply of and demand for such product types.

(ii) The Parties shall use the common logo for the purpose of identifying quali-fied energy-efficient product types listed in Annex C of the energy-efficiency labelling programme.

(iii) The Parties shall ensure that common specifications encourage continuing improvement in efficiency, taking into account the most advanced technical practices on the market.

(iv) The common specifications strive to represent not more than 25 per cent of models for which data is available at the time the specifications are set whilst also taking other factors into consideration.

(v)   The Parties shall endeavour to ensure that consumers have the opportunity to identify efficient products by finding the label in the market.

Participation in the ENERGY STAR labelling program is open to any manufacturer, vendor or resale agent who wishes to join the ENERGY STAR labelling program and use the ENERGY STAR common logo (see Figure 10.2). The logo is a means of assurance that identifies qualified products that have been tested in their own facilities or by an independent test laboratory and that meet the common specifications in Annex C. The participating organisation may also self-certify their products.

Annex C of the energy-efficiency labelling programme includes definitions for the following equipment:

- computers;
- monitors;

- integrated computer systems;
- printers;
- fax machines;
- mail machines;
- copiers;
- scanners;
- multi-function devices;
- imaging equipment.

Figure 10.2 Internationally recognised common ENERGY STAR logo

Annex C also sets out guidelines; for example, there are two guidelines, A and B, under which a computer can be qualified as ENERGY STAR-rated. The two guidelines have been developed to provide programme participants with the freedom to approach power management and energy efficiency in different ways.

The following types of computer must be qualified under Guideline A:

- computers that are shipped with the capability to be on networks such that they can remain in their low power/sleep mode whilst their network interface adapter retains the ability to respond to network queries;
- computers that are not shipped with a network interface capability;
- computers shipped to a non-networked environment.

Computers that are shipped with the capability to be on networks that currently require the computer's processor and/or memory to be involved in maintaining its network connection whilst in sleep mode can be qualified under Guideline B.

Computers qualifying under Guideline B are expected to maintain identical network functionality in and out of sleep mode.

Annex C also provides guidelines for equipment testing; the categories are:

- testing conditions;
- testing equipment;
- testing methodology;
- testing documentation.

Since its inception in 1992, the ENERGY STAR programme has overcome many market barriers and helped revolutionise the marketplace for cost-effective, energy-efficient products and services. The program is a trusted source of unbiased information that helps homeowners, businesses and other consumers understand their opportunities for energy savings and to identify the reliable, cost-effective, efficient products and services that capture these savings. The ENERGY STAR program focuses on driving greater efficiency by bringing to market new energy-efficient products that operate well beyond government minimum efficiency requirements across more than 60 product categories for the home and office.

EXTENDING THE LIFESPAN OF COMPUTERS

To ensure the achievement of ongoing environmental Green IT targets, organisations are encouraged to consider extending the life cycle of their ICT equipment. By delaying the replacement of ICT equipment, organisations can reduce the environmental impacts associated with the manufacturing, transport and waste disposal of ICT, electronic and electrical equipment. However, such an initiative will have to be planned carefully. There may be a danger of continuing to use legacy equipment that is less energy-efficient than its potential replacement. The organisation will need to calculate the difference between embodied carbon emissions and carbon emissions associated with power consumption, and calculate the optimum replacement date.

For an accurate assessment, there are many other factors that need to be considered when making a decision to replace or continue with infrastructure and hardware. As an example, Table 10.1 demonstrates the factors pertinent to the replacement of a PC, measured in carbon emission 'points'. Table 10.1 is not intended to be accurate or based on scientific fact, but it does demonstrate the principle that the CO_2 emissions associated with a PC will continue to rise during its lifetime. The final factor for consideration is the power consumption attributed to the PC over its lifetime. As has already been articulated, older PCs may be less energy-efficient, so carbon emissions will increase at an exponential rate.

Therefore, it may actually be more beneficial to replace an older PC model that is less energy-efficient with a newer ENERGY STAR 4.0-rated PC. However, if the newer PC has a high level of embodied carbon emissions associated with it (perhaps it does not conform to EPEAT standards), then it may be more

Table 10.1 Factors pertinent to the replacement of a PC, measured in carbon emission 'points'

Action	Year 1 CO_2	Year 2 CO_2	Year 3 CO_2	Year 4 CO_2
PC purchase	****			
PC build at workshop	****			
PC delivery	****			
End user training	****			
Software upgrades	*	**	**	***
Service desk support	*	**	**	*****
Desktop support	*	**	**	*****
Hardware repair and replacement	*	**	**	****
Additional upgrades	*	**	**	****
PC disposal				****
Total carbon emissions points	21	10	10	25

beneficial to leave the existing hardware on the infrastructure for the foreseeable future. For this reason, any organisation considering buying new, or replacing existing ICT infrastructure, should do so using guidelines and recommendations provided by both ENERGY STAR and EPEAT.

If a decision has been made to replace old ICT infrastructure with new, then, rather than dump the old infrastructure into landfill or other means of disposal, organisations should consider giving the infrastructure a 'second life'. This can be achieved by selling it on the used infrastructure market or donating the kit to schools and charities. However, extreme care and due diligence needs to be applied when considering selling on the unwanted or redundant infrastructure. Recent studies and news articles have found that discarded computers, even those with 'erased' disk drives, may still contain confidential information such as credit card details, financial and banking records, medical records and personal email. Not only is this a potential security breach for the organisation or individual concerned, but it can also be a source of considerable embarrassment.

With the growth of online auctioning sites, the potential for data loss is huge. Many organisations are finding themselves the subject of unwanted media attention due to legacy infrastructure being sold on to other organisations and individuals. For example, in September 2008, *Computer Weekly* reported that, for 99p, an eBay buyer got access to a West Yorkshire council's network using a second-hand virtual private network (VPN) server.[27] The server, previously used by council staff to allow secure and remote connections to the council's network, was bought from a commercial security firm. When the buyer plugged in the Cisco device and switched it on, he was automatically connected to the internal network of the council.

On powering up the hardware, the buyer had expected the device to need network settings to be input, but, without prompting, it connected to the last place it was used, potentially allowing the buyer to explore the council's network. The council is believed to have disposed of the hardware through a hardware recycling firm, without first restoring the factory settings of the device to wipe away the previous connection data.

Therefore, before you donate or dispose of your unwanted infrastructure, you should ensure that it has been wiped clean of all data and potentially harmful information. Just deleting the 'My documents' folder is not good enough; you may also need to delete the web cache, cookies, internet history, email contacts and messages, and any other personal documents that may reside on the kit. Even formatting does not properly sanitise a hard disk. Therefore, the best approach is to use a disk cleaning utility that overwrites data with random data so that it becomes unrecoverable.

However, if you are a large or medium-sized organisation, ensuring this level of diligence is not easy. Even in a year, hundreds or even thousands of pieces of infrastructure may be replaced, and for every piece of infrastructure that gets either 'moved on' or disposed of, assurances have to be made that all sensitive data and information has been securely removed or encrypted. It is therefore recommended that organisations use a reputable partner to take responsibility for this important and essential service.

As a minimum a specialist electronic and electrical goods disposal service provider must ensure that secure and comprehensive data destruction takes place. Also, once the piece of infrastructure or asset has been moved on, the third party must continue to monitor its usage and eventual destruction. As we have already established, the asset may very well end up abroad and it will be the responsibility of the third party to ensure that the organisation's data and information doesn't go with it.

Ideally, rather than send the asset to another country or continent that may be using dangerous and exploitative working practices, use a third party that guarantees that the resource and component extraction is carried out exclusively by them. Finally, ensure that the third party can guarantee the non-export of discarded equipment and that all disposals of ICT assets will comply with the WEEE directive.

SET UP ELECTRONIC EQUIPMENT DISPOSAL POINTS FOR EMPLOYEES

For companies and organisations, the use of a specialist third party for ICT disposal is essential and achievable. For individual consumers, however, the opportunity to use a third party is not a realistic option. Therefore companies and organisations can provide an essential service for their employees by setting up electronic equipment disposal points.

The advantage for the participating company or organisation is that they are proving the validity of their Green credentials to their employees, shareholders and customers and, of course, providing an essential and worthwhile service.

The advantage for the employee is that they do not have to use the local authority disposal amenities (most of which tend to reside on the outskirts of the main urban centres) or manoeuvre through complex legislation or local government processes. The typical e-waste collection and reuse/recycling system available to the general public is a multi-faceted process that involves transporting hardware from the individual's home to the appropriate recycling facility. However, for many individual households a collection facility is not provided, so the individual will have to take their electronic and electrical waste to a local or municipal waste centre, from where it will be transported to another central location to be bundled up with other similar electronic products. From the central location, the batches of unwanted electronic and electrical equipment will be transported to a company or organisation that specialises in electronic and electrical waste disposal, where it will be checked for resale value and, if appropriate, resold. If reuse is not an option, the device then enters the recycling process. Specifically, it is disassembled by hand and/or shredded by machine to separate different materials. Parts such as steel, copper, aluminium, glass, plastic will then be used in the manufacture of other electronic and non-electronic goods.

Obviously, for the initiative to work, there are a number of hurdles to be crossed for both parties. First, the initiative will need to be financed and both parties may need to agree a value for the assets, or compensation being provided to the employer for providing the facility. From the employees' point of view, ensuring the security of their personal data will be of the utmost importance. However, there are numerous free software packages available to ensure that complete data removal happens before the employer hands over their ICT asset.

11 IMPROVING YOUR ORGANISATION'S GREEN IT CREDENTIALS

In this section we are going to take a look at the types of initiative you as an individual or your organisation could consider when moving towards being a Green enabler. In many organisations IT is considered to be a 'necessary evil' and a common accusation of IT is that 'we don't really know what you do, but what we do know is that you cost us lots of money!' Being at the forefront of an organisation's Green initiative can go a long way towards changing that perception and therefore helping IT to be seen as a strategic asset to the business.

Delivering Green IT into your organisation cannot just rely on complex data centre rationalisation programmes; there are many other ways in which an organisation can deliver Green IT on a practical level. However, there is nothing stopping ICT service providers from being innovative in their approach. We have already seen how Google has patented floating data centres but other organisations are also taking a lead.

Microsoft has investigated building a data centre in Siberia where low ambient temperatures can help to reduce the energy needed to cool the building and the heat-creating equipment it contains. Sun Microsystems is developing a data centre in an abandoned coalmine in Japan, using water from the ground as a coolant. The annualised savings in electricity costs are estimated to be in excess of £5.5 million. Other organisations including Chicago City Hall are installing 'Green Roofing'.[28] As part of a US Environmental Protection Agency study and initiative to combat urban heat island effect and to improve urban air quality, Mayor Richard M. Daley and the City of Chicago began construction of a 38,800 square foot (total roof area) Green Roof in April 2000. It was completed at the end of 2001 at a cost $2.5 million. Encompassing one square block and 12 stories high, this retrofit application serves as a demonstration project and test Green Roof. As well as providing environmental benefits, the Green Roof provides considerable savings on utility bills.

The benefits derived from Green Roofing include:

- reduction of Urban Heat Island Effect (UHIE) – research at the Tyndale Centre for climate change suggests that there needs to be a 10 per cent increase in Green space in our cities to combat climate change. Green roofs are recognised to have a positive effect on reducing the UHIE;

- biodiversity – Green roofs can provide an important refuge for wildlife in urban areas. Research in Switzerland and the UK has demonstrated that Green roofs can also provide important refuges for rare invertebrate populations;

- water – Green roofs can significantly reduce the surface run-off volumes and rates of rainfall leaving roofs. As a source control mechanism in the sustainable urban drainage system, Green roofs can help to reduce flash floods as a consequence of intense rainfall events. This will become increasingly important as a consequence of climate change. Green roofs also improve the quality of water and although the amount of water is reduced, it is possible to harvest rainfall from roofs that have been Greened;

- thermal performance – Green roofs have been shown to significantly reduce the need for air conditioning in summer and can provide a degree of insulation in winter;

- sound insulation – the combination of soil, plants and trapped layers of air within Green roof systems can act as a sound insulation barrier. Sound waves are absorbed, reflected or deflected. The growing medium tends to block lower sound frequencies whilst the plants block higher frequencies;

- protection of waterproofing – the first Green roofs (in Germany) originated from the practice of covering wet bitumen with approximately 6 cm of sand, intended to protect the wet bitumen from fire. The sand often (accidentally but beneficially) became vegetated. Green roofs have now been shown to double if not triple the life of waterproofing membranes beneath the Green roof;

- air quality – airborne particles and pollutants are filtered from the atmosphere by the substrates and vegetation on a Green roof;

- amenity space – in dense urban environments there is often a lack of Green space for residents. Roof gardens and rooftop parks provide important Green spaces to improve the quality of life for urban residents;

- urban agriculture – roofs, where strong enough, provide a space for urban food growing. Although many large flat roofs may not have the loading capabilities to support food growing, some roofs will and the many balconies in urban areas are ideal.

Whilst the ICT industry plays a vital role in providing innovative and environmentally friendly solutions aimed at tackling the bigger causes of GHG emissions, we can also look to the ICT industry to provide day-to day, simple but effective IT solutions. These may include but are not restricted to the following.

ENVIRONMENTALLY FRIENDLY PRINTING

For most organisations that have grown organically and without any identifiable IT growth plan, there have traditionally been a number of different printing devices available to employees. Depending on location, business unit or just user requirement, anything from a single black and white multifunction copying machine to multiple copying machines, standalone colour laser printers and standalone black and white laser printers may actually still be used and supported. Therefore, many organisations have or are considering embarking on a printer rationalisation exercise which effectively has the goal of replacing standalone printers with network-enabled printers.

For these organisations the traditional goal of printer rationalisation is to reduce the number of standalone printers across the company and consolidate printing devices to reduce the associated support issues and costs. Other goals may include improved warranty and reduced purchase spending by entering into an agreement with a single vendor or supplier. Related to this is, of course, the reduction of embodied and consumed emissions by only purchasing units to which multiple users can connect, rather than having a ratio of one printer per user.

As part of the overall Green IT policy of the organisation, you may decide to ban the purchase and business use of ink-jet printers. A 2007 study commissioned by Epson revealed that as much as 60 per cent of the ink contained in an ink-jet cartridge is wasted when printers instruct consumers to replace half full cartridges. On average ink-jet printers provide an ink usage efficiency of just 58 per cent when used for photo printing and only 47 per cent when used for business printing, such as presentations.

However, if you decide to keep your ink-jet and laser-jet printers, then you should consider using recycled printer cartridges as a means of replacement. The benefits of reusing printer cartridges, as opposed to disposing of them, include a reduced chance of the cartridges being thrown into landfill and therefore toxic waste seeping into the soil. It has been estimated that almost 300 million ink cartridges are dumped annually in the USA alone and, considering that typical printer cartridges are made from engineering-grade polymers, they can take as long as 100 years to decompose. Other embodied emission savings related to ink cartridge production include saving natural resources that would otherwise be used to produce virgin cartridges, such as water and oil.

Regardless of the environmental benefits, using recycled printer cartridges will also save you money. The purchase cost will be lower because only the ink is being replaced and not the container and many manufacturers and suppliers offer take-back programmes for used cartridges. With the majority of manufacturers and suppliers, the programme is not brand-specific and therefore is simple to implement. For larger organisations, suppliers may offer a collection service and for individual users and consumers, there is a multitude of options including supermarket recycling points and recycling by mail.

One of the first steps for a printer rationalisation project should be the gathering of information on printers currently in use throughout the organisation. Where possible, it is far easier and quicker to use modern asset discovery tools to provide automated reporting, many of which are available via open source software. Unfortunately, if the printer rationalisation has yet to take place the probability will be that the large majority of printers will not be connected to the network, so there may also be a need for on site manual audits.

Most modern-day asset discovery solutions can provide you with a comprehensive list of the hardware and software on your infrastructure. However, there will still be a need for a manual verification exercise not only to sense-check the automated reports but also to identify any standalone devices not identified via automation. This process will involve support technicians or engineers visiting each of the organisation's physical locations and verifying the make, model and serial number of each device. The third and final process that can be used is to send a communication to the users via email asking for them to provide you with the relevant details of any standalone printers they may be using.

Once all the necessary information has been gathered, the next step will be to agree a logical printer rationalisation model. Based on discussions with users, senior management, your procurement officer, the CSR manager and the Green IT Champion, an agreed approach and defined solution has to be documented. As a minimum the goals of the exercise should include:

- elimination of the majority of standalone and USB-based printing solutions;

- elimination of ink-jet-based technology;

- all multi-function devices to offer fax, scanning, copying and email capabilities;

- all multi-function devices to offer 'secure printing' capability. (This is the ability of the printer to not directly print documents sent to it until the user is physically located at the printer and inputs a user code or uses a swipe card, to confirm their identity; this will also help to ensure that unwanted print jobs are kept to a minimum which in turn reduces the amount of waste paper produced);

- all multi-function devices must be centrally manageable via an automated asset management tool.

Finally, you must remember to engage with the users to ensure that they are fully aware of the environmental and cost benefits of the printer rationalisation project. Regular communication is vital if you are to win and maintain the support of the users. You will also have to be prepared to accept that you will have no option but to provide some users with standalone printers due to disabilities requirements etc. However, the exception criteria must be fully understood and must adhere to current health and safety laws; therefore, it is essential that you engage with the CSR manager when agreeing the exception criteria for the whole of the organisation.

To achieve significant reductions in printing-associated emissions, you do not have to just rely on large and complex initiatives such as organisational printer rationalisation. There are many other, quicker and easier solutions that can be considered, for example: setting default Green printing options, including duplexing and greyscale on all printing devices. A printer's duplex setting offers the capability of a copier, fax machine, Multi-Functional Device (MFD) or printer to automatically place images on both sides of an output sheet, without manual manipulation of output as an intermediate step. Examples of this are one-sided to two-sided copying and two-sided to two-sided copying. A product is considered to have automatic duplexing capability only if the model includes all accessories needed to satisfy the above conditions.

Use recycled paper – the quality of recycled paper has improved enormously in recent years with many types now of photo-quality level. Coupled with this, the cost of recycled paper has reduced significantly so it now makes good business and environmental sense to procure recycled printer and copier paper for your organisation. However, not all recycled papers are equal – some are better for the environment than others.

In Britain, the National Association of Paper Merchants (NAPM) uses a classification system to identify the proportion and source of waste fibre used to make recycled paper. Using the letters A to D, each letter is accompanied with a figure indicating the percentage of that source. It is important to understand the differences between these four sources of waste.[29]

A – Mill broke – this is pulp which has been on the paper-making machine but not turned into saleable paper. It has virtually the same properties as virgin wood pulp and has always been reused by paper mills. Consequently, the NAPM scheme does not recognise mill broke paper as recycled if it contains more than 25 per cent mill broke and/or virgin wood pulp. The longer fibres in mill broke are used to increase the strength of a paper.

B – Wood-free unprinted waste – the term 'wood-free' causes the most public confusion, as the paper is anything but 'wood-free'. The name indicates that when originally pulped, the 'woody' lignins in the timber were destroyed in a chemical reaction (part of the environmental problem of conventional paper-making), to produce a higher quality paper. If not removed, the lignins – the inflammable part of the wood – cause paper to yellow and become brittle with age (as, for example, old newspapers do). You can see exactly the same process in the way pine furniture changes colour with age. Category B waste comes mainly from paper mills and converters, for example, reel and guillotine trimmings, and has not been printed or 'used' in the generally accepted sense.

C – Wood-free printed waste – this comes nearer to most people's idea of what recycled paper should be made from, but because it must only include wood-free paper, it tends to come from limited specific sources, for example, scrapped work direct from printers or discarded computer printouts and envelopes from very large companies. Some would argue that the chief sources of category C waste should themselves be using recycled paper, and that recycling this is only a cosmetic solution. On the other hand, some newer, longer fibre is essential in producing high-quality recycled paper, and this is the most appropriate source for making the very best recycled qualities. Ink in category C waste is removed either by a cleaning process which chemically 'floats' it off or by dispersing it throughout the pulp. The latter is environmentally preferable, but makes the paper less white and so to some users less attractive.

D – Mechanical and unsorted waste – mechanical paper pulp is more realistically named – the wood is pulped by a combination of heat and maceration, but because it retains the lignins, it is only really suitable for short-term uses – for example, newspapers and directories. The lignins present in category D waste can survive into recycled paper made from it, unless the pulp is retreated to remove them. Consequently, some fading of colours or yellowing when the paper is exposed to daylight make recycled paper containing category D waste more suitable for uses such as photocopies which are filed, or for other items with a more limited life. Category D covers almost all domestic waste paper, including that collected at local recycling centres and paper banks, and is largely used in making packaging materials (brown paper and cardboard boxes, for example). Some of this type of waste is used in making the cheaper recycled office copier and similar types of paper.

The waste content of a recycled paper is indicated by the category letter (above), prefixed with a figure showing the percentage of that category in the manufacture of the paper. Therefore, a paper described as 40C/60D contains 40 per cent category C waste and 60 per cent category D waste. Papers containing the largest proportions of categories C and D waste offer the greatest environmental benefits, principally in saving of waste landfill.

Whilst it might appear that the ultimate environmentally beneficial paper would be made from 100 per cent D waste, factors such as suitability for a particular use and also the more subjective perception of paper quality make this more difficult to achieve.

In addition, category D waste can now contain significant amounts of previously recycled paper, and so progressively shortening the average fibre length. Consequently, some fibre from higher up the chain is needed to maintain its strength. However, this should not be seen as justifying the use of proportions as high as 50 per cent virgin pulp in a 'recycled' paper, as produced by some manufacturers. This is simply (at its best) 'environmental tokenism' or (at its worst) Green Wash, and should be treated accordingly.

Stop using paper altogether – in the early 1990s when email was fast gaining popularity, it was often proclaimed that there would eventually never be a need to print at all, and we would all be working in the 'paperless office'. In reality the opposite has happened because now we have more things to print, and there is a natural tendency to have a printed copy of the document or email to read as well as an electronic copy for storage and auditing purposes. However, in recent years we have seen a plethora of document-sharing services, including free services such as Google Docs, which will allow multiple users to view and make changes to docu-ments at the same time without having to print out multiple copies for the group to mark up and make handwritten changes or observations, all of which saves money, time and naturally reduces the impact on the environment of using huge amounts of paper and the associated energy consumed by the related printing.

Other advances in technology with regard to going paperless include the invention of ereaders and ebooks or to give then their formal title 'electronic document readers'; in fact you may very well be using one to read this book. Electronic document readers are devices that are designed primarily for the purpose of reading digital books and periodicals and use e-ink technology to display content to readers. There are many different models available and, from an environmental perspective, you will need to consider the embodied emissions associated with the manufacture of the devices and the consumed emissions associated with power usage and battery life before making a decision on which make and model to purchase.

There is no doubt, however, that the popularity of ereaders is set to increase hugely. In 2007 global revenues of Amazon's Kindle were recorded at £3 million; in 2010 they are expected to rise to £610 million. By 2020 the proportion of book sales forecast to be electronic is estimated at 50 per cent.[30]

Another opportunity for organisations to reduce the amount of printed matter they produce is by using internet-based technologies to bill and invoice their customers. For the everyday consumer, services such as PayPal and Google Checkout allow payments via credit and debit card, and for many companies, their PayPal and Google Checkout accounts are tied directly to their corporate checking accounts for seamless and transaction fee-free payment processing. They both offer a variety of export formats and integrate with popular accounting packages such as QuickBooks and Microsoft Accounting.

ENERGY-EFFICIENT COMPUTING

We have already identified the opportunities for organisations to be more environmentally focused with regard to procurement and purchasing of devices by using methods such as EPEAT. However, there are also numerous, simple and effective means of being more environmentally focused on the day-to-day usage and operations of our devices. One example is the removal of active screensavers. A monitor uses the same power to run a screensaver as to run a working Microsoft Windows application, so consider removing the screensaver altogether. In fact, certain graphic-intensive screensavers can cause the computer to burn twice as much energy than in active mode, and may actually prevent a computer from entering sleep mode.

A well-known service provider recently recounted a story of a customer organisation that was adamant (despite being advised otherwise) that they needed a corporate screensaver uploaded to every single PC in the organisation. The screensaver contained high-definition graphics and photos and was in excess of 10 megabytes in size and took on average 30 min to download over the network. Predictably, within minutes of the screensaver being downloaded, the Service Desk was inundated with calls from users reporting degradation in response times and in some cases they reported that their PC had stopped working altogether. Unfortunately what this led to was an organisation-wide upgrade project where PCs were fitted with additional processor and memory capability, and in certain cases the PCs were replaced altogether.

Screensavers were originally developed to prevent the permanent etching of patterns on older monochrome monitors. Modern display screens do not suffer as much from this problem, but screensavers are still used. If you want to use your screensaver in conjunction with monitor power management, set the screensaver 'wait time' to less than the period of time after which the monitor enters sleep mode.

Another means of saving energy and therefore money is to ensure that your organisation's PCs and monitors have standby settings, and that they are actively used. There are a number of standby options for PCs and monitors including:

- System standby – drops monitor and computer power use down to 1–3 W each and activates PC or monitor 'wake up' in seconds. This could potentially save the organisation £20–70 per PC and monitor annually.

- System hibernate – drops monitor and computer power use down to 1–3 W each and activates PC or monitor 'wake up' in 20+ seconds. This could potentially save the organisation £20 to £70 per PC and monitor annually.

- Turn off monitor – drops monitor power use down to 1–3 W and activates monitor 'wake up' in seconds. This could potentially save the organisation £10–40 per monitor annually.

Each of the standby options reduces the power requirements once the device is inactive. In addition to entering 'standby' status automatically, the product may also enter this mode at a user-set time of day or immediately in response to user manual action, i.e. using the power button. Also, the majority of modern-day PCs have the additional feature of 'waking up' in response to an external prompt, for example, a remote upgrade programme, but for this to happen, the product must maintain network connectivity whilst inactive, waking up only as necessary.

Even if the device is in system standby or hibernate mode, it can still receive important software updates such as new antivirus definitions and Windows security patches. Today, computer software is generally pre-configured to automatically download and apply updates shortly after resuming from system standby or hibernate. For organisations there are numerous ways for network administrators to ensure that software updates are applied, including:

- configure user's devices to apply software patches and updates as soon as the computer becomes available on the network;

- Windows task scheduler can wake up sleeping devices for updates. Scripts distributed via Microsoft active directory can allow for centrally managed 'scheduled tasks';

- with wake-on-LAN activated, a network administrator can wake up sleeping machines at any time in order to perform on-demand software patches or updates.

Another simple consideration is to ensure that PCs and, especially, monitors are turned off at night or during periods of sustained inactivity. A popular myth often recited is that by leaving computers and other devices on, you will use less energy than turning them off and this also makes them last longer. In reality, the small surge of power created when some devices are turned on is vastly smaller than the energy used by running the device when it is not needed.

A further energy-saving initiative being adopted by many organisations is the replacement of traditional desktop PCs with laptops. Whilst there remain functional differences between laptop and desktop PCs, the gaps between price and performance of the two are beginning to recede significantly. In a report by the Energy Saving Trust, it is stated that on average, a desktop PC and monitor will use approximately seven times more energy in a year than a laptop.[31] It also states that procuring a laptop instead of a desktop PC will result in energy savings of around 85 per cent of the potential desktop PC system consumption. It is further claimed that the average savings from the purchase of a laptop rather than a desktop computer will increase considerably over time, due to changes in use and consumption levels. It is important to note that these savings are based upon comparisons of a desktop and LCD monitor against a laptop with no additional monitor. Research is emerging that demonstrates that where laptop users utilise an additional external monitor instead of, or as well as, the laptop monitor, the energy saving would be reduced by around 33 per cent. As a result of the energy saved through the use of laptop PCs over desktop PCs, savings in associated carbon emissions will also be achieved.

As prices become much more competitive and performance converges, sales of desktop PCs are slowing but laptop sales are increasing rapidly. By 2020 it is expected that there will be approximately 22 million laptops and 23 million desktop PCs in the UK. This represents an 80 per cent increase in laptops installed and a 2 per cent increase in PCs installed, against 2006 figures. Desktops are still likely to have strength in the market in the near future where upgradeability is important, as it has historically been more difficult and expensive to upgrade and repair laptops, especially ultra-portable laptops (those of very compact and lightweight design). This is because laptops tend to be more integrated (and less modular in design) to achieve space and power savings. Upgrades in laptops are usually limited to the RAM and hard drive, although sometimes the CPU and video card modules can also be upgraded.

For desktops, with an increase in price/specification, often a rise in idle power consumption is also observed. In contrast, for more expensive, higher specification laptops the trend is often reversed, with the higher priced machines consuming less energy in order to preserve battery life.

In summary, procurement of laptop computers in preference to desktop computers can result in considerable energy savings and related cost and carbon emission savings, especially taking into account the trend in the laptop market towards greater energy efficiency. However, there are other additional factors to consider in the choice between the two PC products. In particular, organisations should consider consumer requirements, such as the portability offered by laptops and the greater upgradeability potential of desktop PCs. Careful consideration should also be given to the energy overheads of external monitors and docking stations where laptops instead of desktop PCs are selected for large-scale use. Organisations should also be aware that, whilst the energy efficiency of products can greatly influence energy consumption, the amount of time that consumers use the equipment can also be very influential on overall energy use. Because laptop computers offer greater flexibility and mobility, they tend to be used more than conventional desktop PCs.

Another technique to assist organisations in achieving energy-efficient computing is sustainable data storage and the removal of files and documents that are no longer required. Many organisations are now using 'de-duplication' technology to identify and remove identical copies of data, documents, spreadsheets and other file types. Data de-duplication solutions will help to lower the disk and/or the bandwidth capacities required by reducing the organisation's capacity requirements. Effectively this means that fewer disks are needed to store the same amount of data. This translates to less bandwidth being required to move and copy that data across the organisation's networks. Beyond these cost reductions, there is also the added benefit of a reduction in storage and network infrastructure which also leads to a reduction in embodied and consumed GHG emissions.

Although the technology behind it can be quite sophisticated, the concept of data de-duplication is simple. Data de-duplication is the process of examining data to identify any redundancy. In the context of back-up data the same data keeps getting backed up over and over again, consuming more storage space and creating a chain of inefficiency. The following example, though simple, illustrates the potential power of de-duplication.

A 2 MB image embedded in a MS Word document and emailed to dozens of people will then probably be backed up by at least 10 people who in turn will use that image and embed it in other documents. In fact, the image is likely to be proliferated throughout the organisation to the point where it has been embedded in 200 other different documents. This creates 400 MB of additional capacity. With data de-duplication, only one copy of the image is stored, saving 400 MB that would otherwise be utilised.

Removal of software bloat – 'software bloat' is a term used to describe computer programs that have meaningless and unnecessary features that are surplus to requirements for most users, or software that uses far more capacity than it really should do, whilst offering little or no benefit to its users. In fact this practice has led to some industry commentators using the Pareto principle (also known as the 80–20 rule) to suggest that in most cases 80 per cent of users only use 20 per cent of the software's features. In the early days of computing, when computer programmers were severely limited by hard-disk and memory capabilities, software code had to be written with a high degree of discipline to optimise every single byte of capacity. In today's computing world, this is perceived as not being an issue any longer due to the huge amounts of capacity now available. As a result of this, many modern-day computer programs are written without any checks or any degree of discipline applied. Unfortunately this can lead to a vicious circle of poorly written programs that (when released) contain bugs and features that nobody wants, being replaced by upgrades that remove the bugs and features but ultimately raise the footprint of the program by adding more bugs and features.

However, in many cases the programmers are under extreme commercial pressure to get the latest program or application ready for release, and each release must contain something new and exciting to replace the old one. Exacerbating the situation is the fact that programs must now come in an array of formats and compatibility, for example, a simple text document now has to have the option of being saved as a CSV, PDF, DOC or RTF.

When considering the growth of software bloat, Moore's Law is as good an indicator as any. Moore's Law is based on a statement made by Intel co-founder Gordon Moore in 1965. In its simplest form it states that the number of transistors capable of being included on a chip doubles every 24 months. This has been the guiding principle of the high-tech industry ever since the term was coined. It is used to predict technological progress and explains why the computer industry has been able consistently to manufacture hardware products that are smaller, more powerful and less expensive than their predecessors – a dynamic curve that other industries cannot match. This then leads directly to program writers producing ever evolving and larger software and applications.

From an environmental perspective, this has led to a significant increase in the manufacturing of hardware to keep up with the demands of the advances in software, and the need for organisations and consumers continually to upgrade their infrastructure. However, there is hope on the horizon, as demonstrated by Microsoft. With the release of MS Windows 7, Microsoft has provided an operating system that actually slows down a trend of exponential growth first seen with the release of MS Windows 95, as demonstrated in Table 11.1.

Table 11.1 Comparison of minimum hardware requirements for 32-bit Microsoft Windows operating systems

Windows version	Processor	Memory	Hard disk
Windows 95	25 MHz	4 MB	~50 MB
Windows 98	66 MHz	16 MB	~200 MB
Windows 2000	133 MHz	32 MB	650 MB
Windows XP	233 MHz	64 MB	1.5 GB
Windows Vista	800 MHz	512 MB	15 MB
Windows 7	1 GHz	1 GB	16 GB

Moving forward, there is a need to prevent software bloat by ensuring that programmers are more disciplined in their approach to code-writing. This is not going to be easy as the modern-day programmer works in a culture of individualism and is quite happy to demonstrate unique skills, regardless of the consequences. There also needs to be pressure exerted on the manufacturers to remove unnecessary features and 'add-ons' from their products and to remove or make unavailable superfluous features not specifically tailored to the end user that are installed by default.

SUSTAINABLE WORKING PRACTICES

An appropriate mantra for any organisation committed to Green IT should be 'reduce, reuse, recycle'. Organisations should use the Green IT programme as an ideal opportunity to reduce the amount of legacy IT equipment that is currently installed on their infrastructure. Replacement of older PCs with energy-efficient (gold EPEAT rating) PCs should be considered, and if this is not financially or logistically viable, at least replace any old CRT terminals that are still in use with far more efficient LCD models. In recent tests, the average energy usage of a traditional 20-in CRT VGA monitor was 63 per cent higher than a 20-in widescreen LCD monitor.[32] Another option for organisations may be to upgrade from a cold cathode fluorescent lamp (CCFL) monitor to a light-emitting diode (LED) monitor.

CCFL and LED refer to the types of backlight in an LCD monitor. Most LCDs use CCFL backlights, which are not as efficient at filling a screen with light as an LED backlight. CCFL backlights consist of several tubes stacked horizontally across the back of the monitor's panel. With LED backlights, there are many individual LEDs all over the back of the screen that can each be turned off or on. This gives LED displays much more precise control over the amount of light coming through the screen, and they are therefore more efficient at energy consumption. LED-based LCDs also have the potential to perform better than CCFL monitors in colour accuracy and can be manufactured with much thinner panels than a CCFL-based display. Of course, advances in technology provide more and more alternative methods of computing. Apple, for example, have recently released the Apple iPad, which, through its touch-screen technology, removes the need for any peripherals at all, including dedicated standalone screens.

Another consideration for organisations is to reduce the number of PCs that exist on their infrastructure by ensuring that they only provide one PC per user. In fact, many organisations are implementing PC sharing schemes to reduce the PC estate to less than one PC per user. This initiative is becoming more viable and accepted, especially as modern workforces nowadays tend to be more fluid than static, and job-sharing schemes become more popular. Away from the workplace, users are already used to sharing PCs, for example when visiting libraries and internet cafés. It is essential, of course, that security issues are identified and addressed. Whenever an individual uses a computer, its cookies, browser history and other settings save the information that has been accessed. It is therefore essential that all the files and settings are deleted once the user logs off, and before the next user accesses the machine. To ensure that the user's data and files are kept safe, a robust password and data access process also needs to be implemented.

There are many different ways in which an organisation can reuse unwanted, retired or legacy infrastructure. A best practice Asset and Configuration Management process will assist an organisation in identifying where assets can be reassigned and reused. As we have already discussed earlier in the book, a popular practice is the passing on of unwanted infrastructure to schools or charitable organisations. This is an initiative that organisations may wish to manage themselves or carry out using a specialist third party, for example, Computer Aid.[33]

In addition to reusing infrastructure, organisations should also consider recycling the remaining infrastructure that cannot be reused. To enable more effective and efficient recycling, organisations should only source electrical products that are designed so that they can be easily disassembled to component level, using universally available tools. The ultimate aim for any organisation is to ensure that no electronic or electrical waste is disposed of into a landfill site.

VIRTUALISATION TO SUPPORT GREEN IT

At its simplest level, virtualisation allows you to have two or more computers, running two or more completely different environments, on one piece of hardware. For example, with virtualisation you can have two different operating systems on one system; alternatively, you could host a MS Windows XP desktop and a MS Windows Vista desktop on one workstation. Virtualisation essentially decouples users and applications from the specific hardware characteristics of the systems they use. The benefits of the flexibility provided by virtualisation include:

- simplified system upgrades by allowing capture of the state of a virtual machine and then transporting that state in its entirety from an old to a new host system;

- enabling computing resources such as processor, memory and storage to be delivered dynamically (only used when needed) rather than delivered in fixed amounts;

- elimination of infrastructure sprawl through the deployment of virtual servers and storage that run securely across a shared hardware environment;

- allowing organisations to increase server utilisation rates and simplify their infrastructure;

- delivery of lower management costs;

- the reduction in energy consumption and therefore a reduced or lower organisational carbon footprint;

- creating 'virtual servers' instead of procuring physical new servers.

In association with or as an alternative to virtualisation, data centre rationalisation can also deliver reduced energy usage. A typical data centre rationalisation project can take an organisation from having multiple legacy data centres to smaller and more modern and energy efficient data centres. It gives the organisation the opportunity to remove a large number of servers with low utilisation and replace them with more compact servers with more powerful and higher utilisation capability. At the same time inefficient legacy servers without power management can also be replaced or retired. The cost and emissions savings that can be realised are potentially extensive considering that a server requires the same amount of power to cool it as to run it and servers are typically only loaded to 30 or 40 per cent of capacity.

DATA CENTRE MANAGEMENT AND IMPROVEMENT

IT data centres are potentially the largest contributor to an organisation's carbon footprint and must become more efficient if organisations are going to reduce their overall environmental impact. There has been a myriad of reported statistics and facts recently, some of which include:

- the world's data centres are said to collectively produce more carbon emissions than whole nations, including Italy, the Netherlands and Argentina;
- the average data centre uses the equivalent electricity consumption of 4,000 homes;
- globally, data centres are responsible for creating approximately 150 million tonnes of carbon annually;
- poor data centre cooling and power usage globally leads to more than 60 million MW hours of wasted energy.

Further research has indicated that electricity consumed in data centres, including enterprise servers, ICT equipment, cooling equipment and power equipment, is expected to contribute substantially to the electricity consumed in the commercial sector in the near future. Western European electricity consumption of data centres was estimated at 56 TWh/year in 2007 and is projected to increase to 104 TWh/year by 2020.[34]

The projected energy consumption rise poses a problem for EU energy and environmental policies. It is important that the energy efficiency of data centres is maximised to ensure that the carbon emissions and other impacts, such as the strain on infrastructure associated with increases in energy consumption, are mitigated.

In March 2007 the EU Code of Conduct for Data Centres was initiated. This is a voluntary scheme within the EU that provides a platform to bring together European data centre owners and operators, data centre equipment and component manufacturers, service providers and other large procurers of such equipment to discuss and agree voluntary actions which will improve energy efficiency. This Code of Conduct (CoC), coordinated by EU Joint Research Centre, proposes general principles and practical actions to be followed by all parties involved in data centres, operating in the EU, to result in more efficient and economic use of energy, without jeopardising the reliability and operational continuity of the services provided by data centres.

The EU CoC for data centres focuses on all buildings, facilities and rooms which contain enterprise servers, server communication equipment, cooling and power equipment and any other hardware and network configurations that provide a form of data service. The CoC covers two main areas of energy consuming equipment in the data centres, IT loads and facilities loads, but ultimately considers the data centre as a complete system and being oriented on the optimisation of the IT system and the infrastructure in order to deliver the desired services in the most efficient manner. The first CoC on Data Centres Energy Efficiency (Version 1.0) from October 2008 came into force at the beginning of 2009.

A key requirement of signing up to the CoC is that interested parties with existing data centres must submit initial energy usage measurements of at least one month's duration, before undertaking an energy audit to identify where savings can be made. The next step is for organisations to submit an action plan, which includes a range of intended best practices. Suggested measures include improving system resource utilisation by employing technologies such as virtualisation as well as optimising the design, configuration and management of energy-hungry cooling systems.

A further commitment is the monitoring of energy consumption on a regular basis and providing the EU's Directorate General Joint Research Centre (DG JRC) with an annual report outlining any improved energy efficiency practices that have been introduced. The DG JRC compares these implemented practices with the promised measures laid out in the action plan and has the right to end an organisation's participation if it believes that progress has been too slow or if members have failed to meet their reporting requirements. Meanwhile, a Data Collection Working Group has also been set up to collect, correlate and analyse information from all contributors in order to work out trends and potentially form the basis of energy efficiency targets in future.

However, the DG JRC has no powers of censure beyond the right to terminate participation, as the code is not mandatory. This means that there is no formal auditing process for compliance beyond the submission of self-certification documentation. There is no accreditation scheme to recognise either membership or compliance.

For a large number of organisations, a complete redesign or refit of their data centres is either financially or logistically unrealistic. Therefore, they may need to consider other initiatives that require low-level investment, or even just a change of working practice. The types of initiative that could be considered include turning off servers outside the service hours dictated by their Service Level Agreements (SLAs). Service providers who purport to follow best practice will review and regularly update their SLAs with their customers. By re-evaluating existing SLAs or implementing them if they do not already exist, organisations can realise potentially significant carbon reductions and financial savings. In many organisations ICT has grown organically without any real planning and therefore infrastructure sprawl is commonplace. This has led to ICT groups supporting old and legacy services, sitting on servers and desktops that remain switched on and run continually 24 × 7, 365 days a year.

A best-practice compliant Service Level Management (SLM) exercise will identify all current services, identify their customers and users and assign a service or product owner to manage them from an IT perspective. Once these actions have been fulfilled, ICT and the customer can start to discuss and negotiate a realistic level of support, availability and reliability for the services.

The outcome of this exercise should lead to a pragmatic and achievable set of agreements that will ensure that services are both fit for purpose and fit for use. In turn, through regular monitoring and reporting, the levels of service can be analysed, discussed and re-evaluated to ensure that the right services are provided at the right times. If managed properly, regular monitoring of services will also identify when a service is no longer required or needs to be retired. This in turn will lead to infrastructure that is no longer required being identified and either reused, recycled or removed permanently.

Whether the infrastructure is being reduced as a result of a virtualisation or a data centre rationalisation project, many organisations will consider how to reuse or recycle the unwanted infrastructure as part of the overall project. One such organisation is Corin Ltd, who, after a rationalisation project, decided to reuse their old infrastructure by building a Disaster Recovery solution.[35] Their Head of IT, David Berwick, explained,

> Recent legislative changes and the environmental trends being adopted by many businesses, mean that IT managers need to be much more aware of environmental issues when planning changes to the data centre. On top of this, when attempting to provide the traditional expectations of being able to deliver greater results for less cost each year, the 2008/9 recession generated much greater challenges for us, with our IT department facing reduced operating expenditure and suspended capital budgets. As a result, plans that had been made for refreshing the infrastructure were either delayed or put on hold as we were expected to tighten our belts for the coming financial challenge.

Berwick further explained,

> These financial pressures, coupled with a drive towards environmental considerations meant things became very difficult going forward, especially for a high consumer of energy such as our organisation. The adoption of a virtualised infrastructure has gone some way to helping us achieve both our environmental and financial goals. One of the main benefits of virtualisation is that the underlying physical architecture is relatively independent of the virtual servers. With the right infrastructure, virtual servers can easily be moved between platforms of different processing ability and whilst some processing power may be reduced by moving to a smaller host, the service can continue to run in a slightly degraded state, almost without notice by the end user.

Berwick continued,

> When we planned the refresh of our production environment, we decided to adopt a standard configuration of our servers to ensure that we could correctly size and evenly distribute the virtualised workloads. After the live services had been migrated across to the new virtual platform, we were left with a number of servers that, although weren't the latest specification, still had the capability of offering significant processing power, with some upgrades to the hardware. As costs were being squeezed, we decided that rather than scrap these servers, we could reuse this equipment to provide the basis of a disaster recovery platform for our virtualised production environment. We re-used many components (CPUs, memory, network cards etc.) from the server farm that we had built up over the years to create a small number of more powerful servers. We then used these servers in a virtual cluster to provide suitable processing power for the Disaster Recovery environment. Rather than provide full systems recovery, we targeted this solution at the services deemed core to the business and although the throughput was lower than in the live environment, it was deemed acceptable as this would only be used when operating in a business continuity environment.

Berwick went on to state that although this was a relatively cost-effective option which worked well for his organisation, there are some notes of caution that other organisations may need to consider. These include ensuring that the processing power of the upgraded system is monitored and carefully analysed. A current processor can potentially operate anywhere from two to five VMs per core. However, the number of physical servers required to replicate an acceptable processing power with the older hardware may be so substantial it renders the proposition unworkable. The service provider also needs to remember that IT components have a finite life and the upgrade option may not be suitable if the manufacturer has discontinued the parts or withdrawn support for key components of ageing systems. The service provider also needs to consider how long they plan to continue to use the legacy hardware, as support issues will become more frequent as the hardware gets older.

In his summary, Berwick stated that the power consumption of the older hardware needs to be factored in to any initiative to reuse it. He went on to say:

> New technology brings advances in the reduction of the power requirements for servers. By extending the life of older less efficient hardware, you could be increasing you future power requirements. If the old hardware is being used in a DR cold start environment then this could be OK, but the expectation for almost instantaneous recovery methods means warm start is the preferred method and this would mean keeping the older inefficient hardware constantly switched on and draining your valuable power resources.

Finally, other 'low hanging fruit' considerations for infrastructure reduction include implementing a multi-tiered storage solution, reducing cooling in the data centre, specifying power conversion efficient power supply units and considera-tion of Power over Ethernet which allows electricity and data to pass over a single cable, thus cutting the amount of infrastructure required, saving costs and using less energy.

CLOUD COMPUTING AND SOFTWARE AS A SERVICE (SAAS)

For many organisations wishing to improve the efficiency of their data centres and their associated Green credentials, there are three main and distinct options.

The first is to design and build a brand new state-of-the-art data centre, incorporating the latest in data centre efficiency design and environmental standards.

The second is to improve, where possible, the existing data centre and consider implementing fairly low-risk and low-cost initiatives such as using energy meters to break down energy usage to the level of components such as server, switch, Storage Area Networks (SANs) and Uninterrupted Power Supplies (UPSs). Organisations will then have the option of using the data to perform component level improvements. Other options include to use CPU throttling on servers, to measure the range of power consumed under a variety of loads, and again use the data to improve capacity performance; and finally to thermal profile to identify hot spots and overcooling.

The third option, which many organisations are now considering adopting, is to outsource the data centre to a specialist third-party supplier. With this option comes the opportunity for organisations to completely rethink their entire IT strategy and consider alternative ways of working, including SaaS and cloud computing.

SaaS is a model of software deployment whereby a provider licenses an application to customers for use as a service on demand. SaaS vendors host the application on their own web servers and download the application to the consumer device. Increased high-speed bandwidth makes it practical to locate infrastructure at other sites and still receive the same levels or improved levels of service. From a home consumer perspective a good example of SaaS is Spotify®.[36] This is a piece of software that allows the user to access media files including music and video remotely but without the need to store the files locally on disks because they are streamed from remote infrastructure.

As Service-Oriented Architecture (SOA) becomes commonplace, in addition to SaaS, 'cloud computing' is becoming more popular, thus untying applications from specific infrastructure. Cloud computing services usually provide common business applications online that are accessed from a web browser whilst the software and data are stored on the servers. According to analysts at the Gartner Group, by 2011, early technology adopters will forgo capital expenditures and will instead purchase 40 per cent of their IT infrastructure as a service. The term

'Cloud Computing' is thought to have derived from the fact that most technology diagrams depict the internet or IP availability as a cloud.

The operational benefits to an organisation adopting a cloud-computing model include increased availability, maximised efficiencies, scalability, flexible pricing and charging models, lower management overheads and lower maintenance costs.

Conceptually, cloud computing is difficult to define and can be interpreted in many different ways; it does not seem to have a standardised definition. However, there is some consistent terminology used when describing cloud computing – words such as 'distributed', 'scalability' and of course 'virtualisation'.

In its simplest form, cloud computing is distributed or utility computing over the internet. This basically means that an organisation's computing services can be hosted by specialist third-party suppliers rather than using 'in-house' local data centres. From the organisation's perspective this means that there is no longer a requirement for organisations to own or manage their own infrastructure, since it is bought as a service and the maintenance of the infrastructure becomes the responsibility of the service provider.

Key to the success of cloud computing and a key driver from a Green IT perspective is the ability for cloud computing to be scalable. Scalability means that resources can be added to or removed from the services depending on the organisation's usage requirements or needs. In traditional computing infrastructures, the organisation is restricted to the amount of computing power available. If, however, that computing power is no longer sufficient, then organisations will have to increase their capital expenditure to buy additional resources, which in turn will lead to depreciation costs, and eventually legacy infrastructure that will cease to be fit for purpose or fit for use. With cloud computing, there are no such limitations or capital costs, as cloud computing can provide unlimited computing power on a pay-as-you-use basis. Another distinct advantage for organisations is that cloud computing removes the need to have dedicated IT server rooms and data centres, with all the expenses and, of course, carbon emissions associated with running them.

One major consideration for organisations contemplating a cloud computing model is the fact that pricing models can be complex and hard to define, and differ considerably from service provider to service provider. For example, the cost may consist of the total price of used hours, used storage and the amount of data transferred; therefore, the total cost of the service may be hard to predict or calculate accurately. Another challenge for the customers is the lack of open standards between cloud computing providers. Each of the service providers will have their own proprietary application programming interfaces, which can lead to vendor lock-in. Other elements to consider include the reliability of the software providers themselves. In what is still an emerging market, how many of the vendors will still be operation in five or ten years' time? Given the nature of IT and the pace at which technologies develop, another consideration will be how long before what is currently an emerging solution becomes superseded? However, this is a challenge that all service provision models face, as developments in IT continue to happen at an unprecedented pace.

Other challenges that need to be considered include, if the organisation decides to transfer its computing to a competing service provider, will it be able to take its existing data with it? Could the organisation lose access to its data if it defaults on payments to the service provider and what is the process for permanent removal and related back-up copies of data if they are no longer required? Purely from a Green IT perspective the greatest challenge will be to ensure that the organisation is not just transferring their own carbon footprint to the third party' footprint.

Perhaps the biggest challenge of all for cloud computing, though, is convincing both public and private sector organisations that it is secure and that their data and information will be protected. Many organisations are extremely nervous about letting a third party have ownership and control of their data and information, and there are also privacy laws such as the Data Protection Act 1998 to which organisations have to adhere.[37] In a cloud-computing model, an organisation's data could potentially be stored and accessed through multiple infrastructures, and will more than likely be sharing that infrastructure with other organisations, some of whom may even be competitors. It is therefore possible that although the data is stored in different virtualised environments, unauthorised individuals might be able to access their data if there is a lack of secure resilience in the infrastructure.

For many organisations it is fair to say that a transition to cloud computing is a huge step to take, from both a technical and a cultural point of view. This is especially true for larger organisations, so perhaps it's only natural that an organisation considers a tactical solution such as server rationalisation to address immediate issues whilst considering third-party cloud computing as a strategic solution for the future. However, the organisation may wish to consider a 'compromise solution' between internal IT service provision and a complete third-party outsource. A solution that some organisations are considering is the 'private cloud' almost as a stepping stone to the public cloud. Essentially, the private cloud is an internally facing version of public cloud computing, which works within the confines of the protected corporate infrastructure of the organisation. It provides the organisation with the same or similar advantages as public cloud computing, but has the additional advantage of adhering to the organisation's own information security governance.

If an organisation decides that it wants to move to a cloud-computing model and use a third party to deliver software as a service, then a set of robust agreements in the form of third-party contracts and SLAs is essential. The next chapter of the book looks at both SLM and supplier management in more detail, since these are the processes most closely associated with the management of such agreements.

REMOTE AND LOCATION-INDEPENDENT WORKING

There are now many different technologies available for organisations who wish to provide the option to their employees of working from home or from satellite locations. These IT remote access solutions include, but are not restricted to, webcams, instant messaging, voiceconferencing and internet video calling solutions such as Skype. The Green benefits for organisations include reduced travel leading to fewer vehicles on the road and therefore reduced carbon emissions, and a reduction in the number of corporate buildings needed, leading to reduced energy requirements, due to less heating and lighting. Another advantage that may not be immediately identifiable is that by providing remote working solutions, staff based at home on a permanent basis can be excluded from needing seats at any recovery arrangements organised with the IT service continuity process. This could lead to reduced recovery requirements which could have a positive effect on CO_2 emissions. However, careful consideration needs to be given before making a decision to embark on this very different method of working. These considerations include:

- what remote alternatives to printing will be provided to the users? You may need to consider high-quality, large flat screens for easier everyday reading and a centralised printing/postal service for large print runs.

- will the organisation allow the use of any existing printers that the users own themselves, and how would these be supported by the IT service provider? It is possible that printing requirements at home will lead to a large-scale procurement of printers leading to a GHG overhead based on both embodied and consumed energy increases.

- if new client equipment is being purchased for the initiative, rather than reusing existing PCs and laptops, will there be a rise in GHG emissions as a result of the extra embodied energy within the new devices?

- will the home-based equipment be portable to allow users to bring their equipment to the office when an incident or service request requires service provider involvement? If not, the need for engineers to go out to visit users may actually lead to an increase in carbon emissions.

- a physical reduction in centralised office space should reduce heating and lighting requirements. However, the organisation needs to ensure that this is offset against the extra heating and lighting costs that users will incur working at home. Also, if there fails to be a proportionate reduction in office accommodation PCs and printers, there is a real danger that both cost and CO_2 efficiency gains will be lost.

- will there still be a need to ensure that some office space and IT equipment is set aside for hot-desking, catering for when home-based staff visit the office?

- there may be an increase in IT-related carbon emissions because of the extra equipment that may be required to support additional resilience of the infrastructure for remote workers.

As the various points above highlight, before organisations decide to commit to providing remote working solutions for their employees, careful planning and consideration need to take place. Organisations will have to seriously consider the health and safety elements of remote working and the social impact of employees working in isolation from their colleagues. Lastly, organisations are going to have to implement the initiative with extreme care and diligence to ensure that they are not simply transferring the organisation's carbon footprint to the individuals working remotely.

PROCESS REDESIGN AND IMPROVEMENT

There are many different methodologies based on best practice that can be used for a myriad of different reasons. Delivering, installing and sustaining Green IT is one of them. The best-known and most widely used methodology is the IT Infrastructure Library commonly referred to as ITIL®. This framework provides process recommendations on how to plan, build, move and operate services to deliver the strategic requirements of the business. Beginning life in 1986 as a result of an Office of Government and Commerce (OGC)[38] project, ITIL has been adopted by organisations worldwide, across all industry sectors and it is this continued success that has seen it evolve into its current third version.

Whilst the framework itself does not explicitly mention environmental factors, the two overarching objectives of ITIL are to deliver quality IT services which are cost effective or cost reducing and to align IT services with the objectives of the organisation as a whole. It does this by enabling the delivery of services to the customer that are deemed to be both fit for purpose and fit for use. ITIL version 3 is centred on five core publications [Strategy, Design, Transition, Operation and Continual Service Improvement (CSI)], each representing a stage in the life cycle of a service. It is this life-cycle approach that ensures that each stage beneficially influences the others, and through constant sets of checks, ensures that the IT organisation can adapt and respond to the changing business need.

From an organisation's standpoint, it is primarily interested in the results of the IT service such as the ability to reach a global market through ecommerce solutions, deliver online services to its customers or be the first to market with new products. This means that it is the IT service provider working within various governance frameworks that will probably be responsible for ensuring that regulatory and legal requirements are included in the design and delivery of the service, including those relating to Green IT. It is therefore essential that Green IT is considered within the complete ITIL life cycle.

Green IT and ITIL service strategy

It is during the service strategy stage that the long-term goals and objectives of the organisation and the service provider are first considered. The key objective of this life-cycle stage is to assist service providers in outperforming the competition. In order to do this, the service provider has to understand what its customers value and to align their service portfolio with this. It is through the portfolio of services that one service provider can distinguish itself from its competitors.[39] It is during the strategy phase that organisations need to consider their supplier strategy, including potential partnerships with organisations from whom they wish to purchase services.

For any organisation embarking on a Green IT journey, it is essential that their suppliers are aligned and fully understand the organisation's cultural, moral and environmental position. This can be ensured by asking questions, tendering documents and through supplier performance management. Whilst Supplier Management as a process sits predominantly in the Service Design element of the life cycle, it is important to recognise that it begins as part of the organisation's strategic decision-making.

The key processes that sit with Service Strategy and align with Green IT include Demand Management. Demand Management is closely linked with understanding the patterns of business activity i.e. when, how often, and what volumes and at what frequency the business uses its ICT services. For example, an organisation looking to implement a remote access solution such as Citrix to provide a slimmed-down infrastructure should consider how many users will connect, where they will connect from, how long they will stay connected for, and whether Citrix is to be the primary solution to access applications.

Armed with this evaluation or projection of business need, the ICT Service Provider can plan how it will use components, and plan procurement of capacity and infrastructure components to avoid the purchase of unnecessary storage, bandwidth, hardware and the facilities to site them and the associated carbon emissions.

Key to effective Demand Management is understanding the organisation's patterns of business activity and user profiles. This will help the service provider to ensure that only the required infrastructure capacity is in place for the agreed services at the agreed times. Doing this will assist the organisation enormously in ensuring that the infrastructure is not oversized and therefore responsible for both excessive and unwarranted embodied and consumed carbon emissions. Looking at the wider implications, Demand Management will have a direct influence on the Capacity Management process, a critical process for Green IT. In turn, Capacity Management modelling and metrics will allow comparisons with predicted and actual usage.

Green IT and ITIL service design

It is during the service design phase that organisations can begin to translate strategy into the reality of service operation, and service design needs to be considered as part of the wider organisational change. Arguably it is the processes in this phase of the life cycle that will have the biggest impact on the organisation's Green IT policy and where it can leverage the most value. For the service design stage to be successful, it is essential that requirements are captured from those individuals and teams that have an understanding of both environmental issues as well as regulatory and legislative requirements. This may include the CSO and Green IT Champion and a proper collaboration must take place between them and the technical experts in the ICT department.

In order to ensure that this happens, the following actions need to be considered:

- capturing the Service Level Requirements (SLRs)[40] – typically these will be captured as part of the SLM process. These requirements must include carbon and other GHG emission reduction targets as well as acceptable power consumption;
- ensuring that the capacity management process models feed into the SLRs to ensure that the new service can be supported within the existing infrastructure or new infrastructure components perform as defined;

- ensuring that a Business Impact Analysis (BIA) and risk assessment exercises are completed as a contribution to the Capacity, Availability and IT Service Continuity management processes;

- involving supplier management if procurement is needed.

The most relevant processes within the Service Design stage of the life cycle, supporting the delivery of the organisation's Green IT aspirations, include SLM. It is the goal of SLM to ensure that an agreed level of service is provided for all current IT services, and that future services are delivered to agreed achievable targets. SLM is arguably the most critical process to Green IT as it supports the opportunity of the organisation and IT to really understand what is needed by both sides and the current capability of IT to deliver it.

The resulting SLA will be a direct result of negotiations between the two sets of parties to arrive at a realistic, achievable and measurable set of targets. However, it will be a waste of time for IT to agree to a set of targets i.e. reduce power consumption of the server estate, if the business is not committed to investing in virtualisation, monitoring tools and agreed downtime. As part of the negotiations, IT may need to ask the organisation some uncomfortable questions, i.e. whether servers really need to be switched on 24/7 or whether this indicates a lack of understanding of either party to the SLA.

It is essential that the SLA includes the responsibilities of both parties to the agreement, and that all interested parties remain committed to them. For example, it is pointless to agree to a file storage limit on users' mailboxes (implemented in support of an overall strategy to reduce storage and therefore the related embodied and consumed CO_2 emissions) if the business overrules the enforcement when an executive shouts loudly. Stakeholder and senior management buy-in is important for the success of all our Green IT targets, but here more than ever it will be essential. SLM plays a vital role in developing the relationship between the business and IT through ongoing communication. It does this not only through the capturing of requirements and negotiation, but also through service review and the alignment of ongoing Service Improvement Programs (SIPs) with the Green IT Action Plan. As a key process in the service design stage, SLM interacts with all the other service design processes, most noticeably, Supplier Management.

Supplier Management has a key role to play in ensuring that the organisation's third-party suppliers are made aware of, and deliver, the Green IT requirements when making an invitation to tender, and ultimately when delivering the service.

From a wider organisational context, if the industry-predicted trend towards utility-based computing becomes a reality, it will be imperative for all organisations to ensure that they align strategically with partners who share the same cultural and environmental goals. Currently, organisations may be selecting suppliers based on their ability to deliver services that adhere to regulations and legislation such as RoHS and the WEEE Directive. However, in the very near future, organisations will increasingly have to entrust aspects of their Green IT policy and action plan (including their CRC responsibilities) to their third-party suppliers. This makes the management of supplier performance and their ability to deliver on their contractual terms vital to the success of Green IT. To assist the organisation in managing their suppliers, ITIL suggests that all information about suppliers, their contractual terms and performance should be stored in a centrally managed logical repository, commonly referred to as the Supplier and Contract Database (SCD).

Another service design process, capacity management, also has relevance through all stages in the service life cycle. The goal of capacity management is to provide cost-justifiable IT capacity that matches the agreed needs of the business in a timely manner. Capacity management can be traced back to the era of mainframe computing where (although simpler to manage centrally) capacity was expensive and difficult to upgrade. Unfortunately, IT seemed to lose the ability of good capacity planning and management with the advent of distributed computing. As organisations are seemingly considering returning to centralised back-end architectures such as cloud computing and thin clients, good capacity management is essential if the organisation is going to remain focused on delivering sustainable and environmentally friendly computing.

Providing a focal point for all performance-related issues for both services and components, capacity management allows an organisation to decide which components to upgrade, i.e. memory, storage or bandwidth, and when to upgrade them. From a Green IT perspective, the organisation will need to be careful not to invest too early in additional capacity that will increase both the embodied and consumed carbon emission of the ICT infrastructure. Nevertheless, the organisation cannot risk poor performance and bottlenecks that will lead to a loss of service and customer dissatisfaction. The IT service provider will need to manage the balance between cost and capacity versus supply and demand, and ultimately the environmental consequences of its decision-making.

A key deliverable of capacity management is the capacity plan which can be used as a business case inviting investment in capacity. The plan should outline existing capacity and the need for additional capacity and will support the mantra of reduce capacity reuse decommissioned components, licences etc. and recycle components that can be considered redundant. This provides the IT service provider with a clear opportunity to exploit the technology expertise that exists within the organisation and to act as an enabler of the organisation's commitment to GHG emissions reduction.

Capacity management is heavily reliant on the information provided by demand management in the service strategy phase. It uses the information provided to influence capacity planning and potentially supports the delivery of Green IT by influencing user behaviour. An example of this is to use the patterns of business activity highlighted by demand management, to implement financial incentives such as differential charging. This is a technique used to smooth out the 'peaks and troughs' of capacity usage by charging different financial rates for the same provision of service, dependent on the time of day the service is being accessed. From a Green IT perspective, this type of technique can lead to organisations considering a 'follow the moon' service provision.

Although still very much at a conceptual stage for most organisations, a follow the moon computing model provides the organisation with the opportunity to draw on the cheapest or cleanest energy supply from anywhere globally, at different times. For larger organisations whose computing operation is global, their data centres can be positioned in different countries and computer processing activities can be switched to wherever the energy supply is cheapest or cleanest at the time. For example, whilst the UK is online, Australia is asleep, so the organisation's computing operation takes place there due to lower energy costs. With the advent of cloud computing, models such as follow the moon computing are expected to become more prevalent in the ICT industry.

Closely linked to capacity management are the processes of availability management and IT Service Continuity Management (ITSCM). The goal of availability management is to ensure that the level of service availability is matched to, or exceeds, the current and future agreed needs of the business, in a cost-effective manner. The success of availability management is dependent on accurate requirements being gathered by service level management, and effective infrastructure design ensuring the requirements are delivered. However, the availability requirements of the organisation with regard to redundancy and resilience may lead to a need to duplicate components, which effectively could lead to increased embodied and consumed emissions associated with the IT infrastructure. It is therefore imperative that an availability plan is produced that will provide guidance and advice on all related issues.

The availability plan will contain information relating to the strategic requirements of the business, so from a Green IT perspective, it will assist the IT service provider in planning any required infrastructure upgrades pragmatically. Linked to the capacity plan, this will ensure that no unnecessary infrastructure is purchased or installed, using unnecessary energy consumption and producing unwanted carbon emissions.

Whereas the focus of availability management is on the day-to-day availability of services, ITSCM will employ similar techniques and solutions, but with a focus on ensuring that the organisation remains up and running in the event of a significant outage or related disaster. The ITSCM process has four key stages; initiation, requirements and strategy, implementation and ongoing operation. It is in the first two stages that the main benefits of Green IT associated with ITSCM can be realised.

The initiation stage covers policy setting, management intention and objectives, terms of reference and scope, including regulatory and customer requirements. At this stage the IT service provider needs to engage with the organisation to ensure that Green IT legislation and regulations are considered in the scope of the ITSCM plan.

The requirements and strategy stage includes arguably some of the most important activities of ITSCM and will enable the IT service provider to identify the Green IT requirements associated with the process. This stage includes two key activities, BIA and risk management. There is a strong complementary role here to that of SLM in capturing requirements.

BIA is how IT discovers the criticality of IT services to the business and the impact of their loss, whereas risk management[41] identifies the risk that could result in a loss of service, and provides the solutions to either removing or mitigating the risk. The results of the BIA and risk management exercises allow the IT service provider to determine the continuity and risk response strategy.

The response strategy will vary between organisations but may typically include improving availability by removing single points of failure, improving the back-up and recovery strategy and improving the environmental infrastructure e.g. by the use of UPS devices. It is also likely that for many organisations, the recovery plans will include the ability to restore a service at an alternative location and for the services to continue to be delivered at a hot standby site or split site with full data mirroring. This of course can potentially lead to a significant increase in the IT service provider's carbon footprint, especially if data centres are being mirrored.

Green IT and ITIL service transition

The Service Transition stage of the life cycle focuses on the transfer of services from Service Design into the live operation, ensuring that services meet the security, availability, capacity and continuity requirements, which have been defined during the Service Strategy phase. Service Transition is also tasked with ensuring that the customer expectations defined earlier in the life cycle are met or exceeded. Key goals include planning and managing the resources needed to establish successfully a new or changed service into production within the predicted cost, quality and time estimates; ensuring that there is minimal unpredicted impact on the production services, operations and support organisation and the provision of clear and comprehensive plans that enable the customer and business change projects to align their activities with the service transition plans.

In short, service transition seeks to transfer new and enhanced services as well as operational and maintenance-related changes into the live environment with minimal surprises. In order to achieve this, there need to be established and repeatable testing strategies, working closely with the modelling and application sizing activities of capacity management where necessary. The two key processes within service transition that will have the widest impact on Green IT are Service Asset and Configuration Management (SACM), and change management.

SACM is considered to be an essential process supporting and enabling all other service management processes throughout the life cycle and is essential to fulfilling the Green IT requirements of the organisation. Within SACM, asset management is concerned with identifying assets in the organisation, maintaining asset inventories (primarily for financial accounting and reporting), whereas configuration management takes a more holistic view. The essential distinction between the two processes is that configuration management creates a logical model of services by mapping the relationships between Configuration Items (CIs),[42] infrastructure, assets and services. It is considered mandatory to record the status of a CI as part of the status accounting stage, to enable life-cycle reporting.

It is the logical model and status accounting that gives confidence in all Service Management activities, from capturing baselines for testing and deployment, assessing the risk of making changes, determining the impact of incidents, providing the basis for a technical service catalogue and for financial planning. As part of the IT service provider's commitment to reducing the organisation's carbon footprint, it is essential for them to know what infrastructure exists in the organisation's testing, production and replica sites. This is key to ensuring the management and procurement of additional components only when absolutely required, the reuse of decommissioned CIs to guarantee that they have a longer lifespan, and to ensure that when the CIs are eventually disposed of, there is a confidence that they are done so in accordance with the WEEE Directive.

Despite the essential nature of SACM, it is rarely fully implemented in organisations as it is perceived to take too long to implement and maintaining the Configuration Management System (CMS) is seen as a significant overhead.[43] However, without a CMS which identifies at the very least our hardware and infrastructure CIs, their performance characteristics and any environmental incidents and changes relating to them, it becomes far more difficult to achieve the Green IT objectives.

The change management process is a key process not only in service transition, but in the ITIL framework overall.[44] Change management is responsible for the success of a change or release by ensuring that the change is recorded, assessed, authorised, scheduled, built, tested, implemented and reviewed properly. As part of the assessment of changes, ITIL advocates the consideration of 7 Rs:

- What is the reason for the change?
- Who raised it?
- What are the relationships to other changes?
- What is the expected return?
- Who will be responsible for its success?
- What are the risks of making the change?
- What resources will be needed?

At each stage of the change assessment, the environmental impact of the change needs to be considered. Change management cannot approve changes in isolation and may require a group of experts and interested parties to fully consider the 7 Rs. The Change Advisory Board (CAB) will therefore need to consider including a Green IT representative, perhaps the Green IT Champion, when considering major changes to infrastructure, power requirements, and data centre design or suppliers.[45]

Green IT and ITIL service operation
It is during the service operation phase of the life cycle that the services planned, designed and made live in the previous stages are deployed live into production. It is the purpose of service operation to manage and maintain the technology, and coordinate and carry out the activities and processes required to deliver the

service outcomes to the organisation. It is the only stage of the life cycle that incorporates specific functional groups.[46]

These are defined as:

(i) The service desk:

- the single point of contact between the users and IT;
- most typically engaged in managing incidents and handling requests from users;

(ii) technical management:

- a logical function comprising groups of experts and custodians of technical knowledge relating to infrastructure;

(iii) applications management:

- a logical function comprising all groups of application experts, again considered to be custodians of technical knowledge relating to applications;

(iv) IT operations management, responsibilities divided between two sub-functions:

- operations control – focuses on day-to-day repeatable activities including back-ups, file and print, batch processing and monitoring the environment;
- facilities management – focuses on the physical and environmental aspects of the service, including data centres, power consumption, rack space, floor load and temperature control and cooling.

From a Green IT perspective, it is the expert groups that will assist the organisation in realising the physical changes that will need to take place to deliver more environmentally friendly infrastructure and services. The organisation will need to harness the expertise of technical and application management at all stages of the life cycle. The success of Green IT objectives will depend on targets being translated into appropriate thresholds and meaningful measurements being captured, responded to and reported on in service operation. Essentially, service operation is the 'doing' bit of IT, and without mature operational activities monitoring CIs and recording their performance, there will be no metrics to support capacity and availability management or our CSI initiatives.

Service operation is also responsible for managing five processes, including:

- event management;
- incident management;
- problem management;

- request fulfilment;
- access management.

From a Green IT perspective the two most influential are event management and incident management. Event management is the process that provides the ability to detect, interpret and initiate appropriate action for events. It is the basis for operational monitoring and control and is the entry point for many service operation activities. ITIL defines an event as any occurrence of significance to the management of services including infrastructure and applications. IT needs to define events and the required response to them as part of service design and respond to them in service operation, typically as part of operations control within the operations bridge.[47]

Event types are defined as:

- informational – typically an expected or graceful event e.g. back-up completed successfully, or temperature within expected parameters. The response here may be to log the event for information or future analysis;
- warning – a threshold has been breached which may serve as an early warning of threat of service loss e.g. a disk has breached an expected usage limit or power consumption is rising;
- exceptions – indicates a failure or catastrophe of some sort, e.g. a disk is full or air conditioning has failed.

Any event requiring some kind of action to be taken, usually human, is referred to as an alert. From a Green IT perspective, alerts will be the basis for remedial actions including response to power increases and temperature changes that fall outside our defined Power Utilisation Efficiency (PUE) in the data centre.[48]

Incident management is a reactive process where the goal is to resolve incidents as quickly as possible to ensure continued service delivery. An incident is defined as any event which is not part of the normal operation of a service and which causes or may cause disruption to a service. The introduction of Green IT-related targets is likely to require additional incident categories and priorities to be added to our service management tools. There will also be a requirement to educate all support staff to enable appropriate handling of environmental incidents and enable effective reporting of them.

ITIL defines the relative importance of incidents by priority and suggests that priority should be defined by both the impact and the urgency of the incident. Impact is concerned with the actual or potential effect of the incident and urgency as a time measurement relating to when the impact may be felt or how long we have to resolve the incident. An example of an environmental impact may be potential breach of our CRC commitments by exceeding defined power consumption or temperature thresholds. An environmental incident may be related to a warning or exception event and where possible the goals of incident management should be supported through automated logging and management via integration with event management tools.

Green IT and ITIL Continual Service Improvement (CSI)

The primary goal of CSI is to continually align and realign IT services to the changing needs of the business by identifying and implementing improvements to services, processes and activities. CSI advocates that improvement is a goal in itself and should not simply be a response to a failure or loss. It focuses on the improvement of service management as a whole, the continual alignment of the service portfolio with changing organisational need, and ensuring that processes, services and activities are improved.

Implementing Green IT policies and action plans is likely to be a long-term and ongoing project within an organisation and to fall within the scope of CSI. CSI is concerned with meeting the changing needs of the organisation, and meeting Green IT legislation, reducing power consumption and carbon emissions are all valid drivers. At the very least, the success of the Green IT Policy and associated action plans can be enhanced with CSI principles and techniques.

To enable these goals, it is essential to have a clear understanding of how services, processes and all life-cycle stages are performing. It is essential therefore to have relevant metrics defined and captured, focusing on technology, services and processes. CSI introduces a number of tools used to ensure that improvement opportunities are identified and implemented effectively, including benchmarking, balanced scorecards and the use of metrics.

The first of these tools and the underpinning basis for CSI is the Deming cycle, a continuous quality improvement model consisting of a logical sequence of four repetitive steps for continuous improvement and learning: plan, do, check (or study) and act.[49] It will be essential to the ongoing success of Green IT to take a structured approach to its implementation and ensure the audit of our short-term goals before moving on to implement more costly long-term plans.

Another of the CSI tools available is the 7-Step Improvement Process. This seeks to translate the vision of the organisation through strategic, tactical and operational goals. If, for example, our goal is to turn the CRC into a business opportunity, it would help to define what we should measure and what we can measure in the form of data through its processing and analysis, and to identify who and in what form it should be presented to ensure implementation of corrective actions.

12 CONCLUSION

Whilst there can be little doubt that Green IT makes sense for each and every stakeholder, what is undeniable is that there will also be a series of potential issues and challenges that will need to be overcome. Many of these challenges have already been identified in this book, as have the suggestions on how they can be overcome. Whatever the challenges, what is absolutely vital to making Green IT work in the organisation is senior management buy-in at the highest possible level and having an enthusiastic stakeholder group that wants to be involved and is willing to remain focused, even when some of the many initiatives become increasingly difficult to implement. Keeping the momentum going and enthusing the organisation as a whole is absolutely key to Green IT being a success, and this will only happen if a dedicated and driven Green IT Champion (or Champions) is in place and given the time and resources to deliver.

Another challenge is to manage the increasingly complex interfaces and relationships associated with Green IT. This becomes especially difficult if the organisation is to a large extent reliant on outsource partners or third-party suppliers. Robust Business Relationship Management and Supplier Management processes need to be in place, and there will be an increasing need for environmental and Green requirements to be articulated in third-party Invitations To Tender (ITT).

It is also worthwhile remembering the complexities of the ICT infrastructure that must be continually supported whilst a move to a Greener solution is being considered. Achieving a balance between maintaining a stable production environment and being responsive to the changes needed to deliver Green IT can potentially be very challenging, even for a small to medium-sized enterprise. Nowadays, for the majority of organisations, the infrastructure tends to be distributed or virtual, so even considering a printer rationalisation programme can bring with it enormous challenges. Ideally in these situations, the needs of Green IT have to be considered in the context of wider organisational change. Therefore, whilst the advantages of reducing the number of standalone printers are evident from an environmental perspective, there also needs to be identification of wider business benefits, for example, reduced cost associated with replacing ink cartridges.

Other challenges include having to ensure that there is a balance between taking a pragmatic approach to Green IT and being completely overwhelmed by the associated bureaucracy and red tape. Fortunately there are a number of organisations who can help both businesses and individuals not only to understand their legal and legislative commitments but also to provide advice on how to address them and implement the changes necessary to deliver their obligations.

Two such UK-based organisations are the Carbon Trust (who have already been introduced earlier in the book) and Global Action Plan.

Global Action Plan has been in existence since 1993 and they run environmental projects in the UK supported by the United Nations Environment Programme (UNEP).[50] Today Global Action Plan assists thousands of people and organisations including schools, businesses, charities and local authorities to make environmental and financial savings by providing advice and solutions for their environmental challenges.[51]

In conjunction with identifying challenges, organisations also need to be aware of the risks associated with Green IT. There is a need to be realistic in our understanding of what can be achieved against reasonable timescales. Risk is an inherent fact of life which goes hand in hand with opportunity, and we all face, evaluate and manage a wide variety of risks throughout both our working and our private lives. Therefore, having a robust risk management process aligned to the Green IT initiative is strongly recommended. Risk management is the discipline of identifying, monitoring and limiting risks. Once the risks have been identified, they must be reduced, removed or managed as acceptable risk.

Some of the risks associated with Green IT include a change in accountabilities and responsibilities for key staff that can lead to a resistance of change and possible resentment towards the initiative and its key supporters. The organisation has to be realistic and recognise that not everyone will be a winner or even benefit from Green IT. For example, as new and modern efficiencies are introduced into the organisation, there may be a case for reducing the size of the workforce; less infrastructure may equal reduced support requirements, and therefore fewer analysts and support engineers.

Another potential risk of implementing Green IT is the accumulation of additional, unplanned costs. It is therefore imperative that proper financial management processes are put in place. This can happen as part of a formal project approach, and if the Green IT Champion does not come from a financial management background, then there will be a need to engage with the organisation's finance directorate. As a minimum there will be an expectation to understand the TCO of the project, and a documented Return on Investment (RIO) will need to be provided.

TCO can be used as a means of calculating the cradle-to-grave costs of a purchase, project or investment. The ICT industry has always been a key exponent of TCO, since there generally tends to be a considerable difference between an IT infrastructure purchase cost and its ongoing operational cost. It is generally accepted that the operational costs of a specific IT component (a server, for example) can be up to eight times that of the hardware and software purchase costs. Good TCO analysis brings out the hidden or non-obvious ownership costs that might otherwise be overlooked in making purchase decisions or planning budgets.

ROI is a performance measure used to evaluate the efficiency of an investment, project or initiative in comparison to other investments, projects or initiatives. To calculate ROI, the return (benefit) of an investment is divided by the cost of the investment; the result is expressed as a percentage or a ratio. Return on Investment is a popular metric because of its versatility and simplicity; that is, if an investment does not have a positive ROI, or if there are other opportunities with a higher ROI, then a decision can be made as to whether the investment should or should not be undertaken.

There are a number of simple steps an organisation can take to identify the potential risks associated with implementing Green IT. The first step is to identify the key assets; this does not necessarily mean just the hardware needed for a server virtualisation project, for example; other key assets may include software, documentation, buildings and people. These can be broken down even further, for example, people 'assets' may relate to users, customers, suppliers and shareholders. Obviously there are some assets that will be of more value to the organisation than others, so an exercise to rate the assets based on the risk impact of the asset being lost or unavailable will also need to take place.

The risk impact rating provides a numeric value matched to the severity of adverse effects, or the magnitude of a loss, caused by the consequences. It is recommended that the numeric rating be represented as the higher the impact, the higher the number. See Table 12.1. Once the assets have been identified and rated, the next step is to identify the types of threat that may affect the assets.

Table 12.1 Risk impact score and description

Impact rating score	Description
1	Minor impact on the business. Inconvenient for staff but unlikely to be noticed by customers. No noticeable financial impact (<£500).
2	Disruption within the business, likely to result in a backlog of work. Quality of service may be impacted. Low financial impact (£500–5,000).
3	Significant impact on the business. The core business processes can still function but certain business areas unable to function. Medium financial impact (£5,000–50,000).
4	Major impact on the business. Critical business process unable to function. High financial impact (>£50,000). Significant stakeholder concern.

Identifying a comprehensive set of threats can be difficult, so it is probably best done by getting a group of interested parties together and holding a simple 'brainstorming' session. The session can be facilitated by the Green IT Champion, and the threats identified will need to be documented. Obvious threats may include loss of service, loss of revenue, a lack of funding, staff alienation, lack of expertise in understanding Green legislation and deliberate circumvention of processes or regulations.

Once the threats have been identified, there will be a need to understand the vulnerability to the threat, by understanding the likelihood of the threat happening. This can be achieved by agreeing a 'likelihood rating' as demonstrated in Table 12.2. To provide a complete understanding of the potential impact of the risk and to prioritise risk mitigation, there will need to be an exercise to multiply the risk impact rating by the vulnerability rating to produce an overall risk rating; see Table 12.3. It is this risk rating that will help the organisation to prioritise the actions needed to either remove or mitigate the risk.

Table 12.2 Likelihood of risk

Likelihood rating	Description
1	Very unlikely to occur within the next 12 months. It is theoretically possible but we do not expect that it will ever occur.
2	Not likely to occur within the next 12 months. There is an outside chance of this occurring, so it cannot be discounted.
3	Likely to occur within the next 12 months. There is a reasonable chance that this could happen.
4	Very likely to occur within the next 12 months, almost a case of 'when' not 'if'.

In some cases a decision may have to be made just to accept the risk. This may be because the cost of mitigating the risk is too high or, even though the impact is high, the likelihood is so low that mitigation cannot be cost-justified. Alternatively, the impact and likelihood are both low, so it is decided that no action will be taken to remove the risk at this time. Finally, the situation might be that the risk has been previously identified and reduced, but some residual risk remains and is accepted.

It is advised that you used a well-recognised risk management methodology when embarking on any risk identification exercise. One of the best-known and most

Table 12.3 Risk vulnerability ratings

Action	Asset	Impact rating score	Threat	Likelihood	Likelihood rating	Vulnerability rating (value of asset x likelihood rating)
Printer rationalisation project	File Print server FTP0002	2	Loss of service due to 500 additional user IDs being created	Likely to occur within next 12 months. There is a reasonable chance that this could happen	3	6
Mobile and remote working project. Operation's employees to be relocated as home workers	Operations Dept	4	Union invoked walkout due to unfair working practices	Not likely to occur within next 12 months. There is an outside chance of this occurring, so it cannot be discounted	2	8

widely used methodologies is the OGC's Management of Risk, commonly referred to as M_o_R. However, there any many other established risk management methodologies, and if you are intending to use an established project management methodology such as PRINCE2®, risk management will be a fundamental component of that methodology.[52]

In conjunction with identifying, managing and removing the risks associated with Green IT, it is also essential to identify the Critical Success Factors (CSFs). CSFs are the elements that are absolutely vital to achieving the objectives, goals and mission of a plan or initiative. The CSFs required for Green IT to be a success include; establishing and maintaining stakeholder 'buy-in', maintaining and managing all the relationships affected or involved in Green IT, integrating with the other business unit's projects and processes that are involved in wider Green initiatives, developing staff with the right knowledge and skills and defining clear accountabilities, roles and responsibilities.

It is no accident that the common theme running through all the CSFs is communication. A well-thought-out and agreed communication plan is a worthwhile starting point to delivering the CSFs. The communication plan for Green IT should deliver a communication strategy that provides a means of checking that communications are well understood, delivered in a timely fashion to the right audience and via the correct medium. For example, the senior management or executive board may need a formal, monthly presentation from the project steering committee, whereas the updates generated for standard organisation-wide communication can be provided via posters, emails, newsletters or web blogs and podcasts. It is also worth remembering that the role of the media in Green IT, the last thing anyone wants to be a part of, is a great success story no one has ever heard of! There are now a multitude of websites, whitepapers and other publications relating to Green IT that you can use to widen your communication and 'spread the word'.

Whilst it is critical to the success of Green IT to have a communication plan that articulates clearly why we are changing, there are other important factors that need to be considered. This includes identifying any new skills or knowledge required to deliver Green IT and getting the right people into the right roles. The organisation may be lucky enough already to have people with the necessary skills in place; conversely there may also be a need to invest in training and development of people.

Therefore, there will be a requirement to have clearly understood and measurable objectives agreed for both individuals and teams, linked to personal development plans.

The effects of making changes to the organisation or to the way individuals conduct themselves in their day-to-day roles must not be underestimated. Even the slightest of changes can bring with it resistance, and a sense of unease or outright avoidance. If the organisation is to be successful in delivering Green IT, it needs to ensure that a well-established approach to change management is considered. John Kotter, who teaches Leadership at Harvard Business School, states that the most successful change programmes go through a series of eight phases that, in total, usually require a considerable length of time.[53]

The eight phases are to:

(i) establish a sense of urgency:

- help others to see the need for change and the importance of acting immediately;
- start collating evidence on the environmental and political damage being caused by current practices;

- highlight the danger of not meeting legislative commitments;
- understand the impact of global legislation;
- gather information on the cost of current IT service provision;
- gather information on the carbon emissions associated with IT service provider;

(ii) pull together the guiding team:

- make sure that there is a powerful group guiding the change – one with leadership skills, a bias for action, credibility, communication ability, authority and good, analytic skills;

(iii) develop the change vision and strategy:

- clarify how the future will be different from the past, and how you can make that future a reality;
- start measuring and analysing against the information and facts established in phase 1;

(iv) communicate for understanding and buy-in:

- make sure that as many others as possible understand and accept the vision and the strategy for delivery of Green IT;
- Green the staff: socialise the Green issue and educate your workforce;

(v) empower others to act:

- remove as many barriers as possible so that those who want to make the vision a reality can do so;
- ensure that the Green IT Champion is not 'micro-managing';

(vi) plan for and produce short-term wins:

- create some visible, unambiguous successes as soon as possible;
- start implementing 'quick wins';

(vii) not let up:

- press harder and faster after the first success. Be relentless with instituting change after change until the vision is a reality;
- develop a CSI plan;

(viii) create a new culture:

- hold on to the new ways of behaving, and make sure they succeed, until they become a part of the very culture of the group;

- incorporate environmental criteria into procurement decisions;

- start assessing your vendors – beware the 'Green wash'.

Ultimately it is the responsibility of whoever is leading the Green IT project to assess the organisation's readiness for change and this can be done by putting together a simple checklist (see Table 12.4). What must not happen as part of the change programme is that one person, i.e. the Green IT Champion, ends up trying to micro-manage everything or develops over-bureaucratic processes that no one will buy into. To conclude, if Green IT is to be successful, the following needs to happen: recognise the challenges and face them, identify the risks and remove or mitigate them, identify the critical success factors that must happen if Green IT is not to fail, build an effective communication plan, and establish a recognised change management framework, to deliver all of the above.

Table 12.4 Assessing the organisation's readiness for change

What to check	Where is the evidence	Comments
Assessment of staff and skill levels	Skills plan	Any skills gaps to be identified and a formal training plan linked to personal development plans to be created.
Is there a documented vision?	Vision/strategy	
Have roles and responsibilities been reviewed?	RACI matrix	
Are specific roles and measures defined?	Performance measures	
Have the skills for each area been defined?	Skills requirements report	Skills requirement report to link into formal training plan.

(Continued)

Table 12.4 *(Continued)*

What to check	Where is the evidence	Comments
Is there an assessment of the personnel skills and requirements?	Assessment report	
Have other areas been considered?	Assessment report	
Have the requirements for the support of the business needs been considered?	Requirements definition	
Has the level risk been documented?	Risk assessment report	Consider appointing the business continuity manager to do this (if one exists).

13 SUMMARY

Green IT is a collection of strategic and tactical initiatives which directly reduce the Carbon Footprint of the organisation's computing operation. Green IT uses the services of IT to help reduce the organisation's overall carbon footprint and to encourage and support Greener behaviour by the organisation's employees, customers and suppliers. Green IT also encourages investment in ICT by the organisation to assist in its own commitments to lowering its carbon footprint. Green IT also ensures that the importance of sustainable resources are understood, by recognising both the consumed and embodied energy associated with ICT resources and uses that understanding to help drive the need for Green IT Policy.

Key Green IT roles include: the Green IT Champion, the CSO, the procurement manager and the corporate social responsibility manager. The key factors driving Green IT are: political, environmental, social and legal. The benefits of Green IT are generally regarded as being: organisational, cost, an improved reputation and the introduction of a sustainable and environmentally aware organisational culture.

Before an organisation can start to deliver the improvements identified in the Green IT plan it first needs to audit its current practices by measuring its current GHG emissions using a recognised carbon footprint calculator and associated methodology. Thereafter, there will be a need to calculate the TCO of Green IT, calculate the pay-off period of the investments in Green IT and understand and document the ROI. A defined set of targets will need to be identified and, importantly, a detailed understanding of legislation and governance that is likely to affect the organisation, either directly or indirectly, is essential.

There are many ways in which an organisation can deliver Green IT improvements, including: environmentally friendly printing, energy-efficient computing, sustainable data storage, virtualisation to support Green IT, cloud computing and SaaS. Other initiatives include: process redesign and improvement using a recognised best practice methodology such as ITIL, remote or mobile working, data centre rationalisation and modernisation, sustainable IT and environmentally friendly asset disposal.

The initiation of any Green IT project will have to be managed using a recognised project management methodology such as PRINCE2 and there must also be recognition that, inevitably, Green IT will bring change to individuals and the organisation as a whole, and will need to be managed accordingly. The different change tools and methodologies we can use include JP Kotters Eight Steps to Effective Change.

It is critical to review at least annually your Green IT plan to accommodate new developments in technology, customer requests, the economy, emerging trends and new or changed legislation. Therefore, it is recommended that the IT Service Provider establishes an improvement action plan, allowing for regular review of progress. The plan will need to allow for measurement of progress against set targets, for example, improved resource efficiency, and reduced air pollution and GHG emissions. The improvement action plan will also act as an enabler for suppliers and other third parties to continue to improve and promote Greener products and services, and ultimately the improvement action plan will provide a means of reporting on the progress and ongoing environmental performance of the ICT service provider.

Finally, ensure that the benefits of Green IT are identified and communicated across the whole of the organisation and to all third-party stakeholders via a well-defined communication plan. The benefits of Green IT include reduced costs, reduced carbon footprint, improved organisational credibility, the introduction of a culture of resourcefulness, adherence to legislation and governance and, of course, the organisation and IT doing their bit to save the planet, and ultimately the human race!

USEFUL REFERENCE SITES

Name	Purpose	Link
BCS (British Computer Society) Data Centre Specialist Group	Organisation	www.dcsg.bcs.org
Carbon Reduction Commitment (UK)	Legislation	www.decc.gov.uk
Carbon Trust	Organisation	www.carbontrust.co.uk
Climate Change Act (UK)	Legislation	www.decc.gov.uk
Climate Group	Organisation	www.theclimategroup.org
Department for Environment, Food & Rural Affairs (Defra) (UK)	Organisation	www.defra.gov.uk
Department of Energy & Climate Change (UK)	Organisation	www.decc.gov.uk
EU Code of Conduct for Data Centres	Voluntary Scheme	www.dcsg.bcs.org
Environment Agency (UK)	Organisation	www.environment-agency.gov.uk
Environment Protection Agency (US)	Organisation	www.epa.gov
Energy Star	Organisation	www.energystar.gov
Energy Saving Trust	Organisation	www.energysavingtrust.org.uk
EPEAT	Voluntary Scheme	www.epeat.net
Gartner Group	Organisation	www.gartner.com
Global e-Sustainability Initiative (GeSI)	Organisation	www.gesi.org
Global Action Plan	Organisation	www.globalactionplan.org.uk
Green Grid	Organisation	www.thegreengrid.org

Green IT Magazine	Publication	www.greenitmagazine.com
Greenpeace	Organisation	www.greenpeace.org.uk
ISO 14000–14001	Standard	www.iso.org/iso/iso_14000_ essentials
Kyoto Protocol	Legislation and Guidance	www. unfccc.int/kyoto_protocol/ items/2830.php
PAS 2050	Voluntary Scheme	www.bsigroup.com/upload/ Standards%20&%20Publications/ Energy/PAS2050.pdf
QA Ltd	Training	www.qa.com
Restriction of Hazardous Substances (RoHS)	Legislation	www.rohs.gov.uk
WEEE Directive	Legislation	www.environment-agency.gov.uk/ business/topics/waste/32084.aspx
SMART 2020	Guidance	www.smart2020.org
Wikipedia	Reference	www.wikipedia.org
World Wide Fund for Nature	Organisation	www.wwf.org.uk

ABBREVIATIONS

AMR	Adaptive Multi-Rate
ANSI	American National Standards Institute
ASA	Advertising Standards Authority
AT&C	Aggregated Technical and Commercial loss
BIA	Business Impact Analysis
BMS	Building Management System
BSI	British Standards Institute
CAB	Change Advisory Board
CCFL	Cold Cathode Fluorescent Lamp
CDM	Clean Development Mechanism
CER	Certified Emission Reductions
CH$_4$	Methane
CI	Configuration Item
CIO	Chief Information Officer
CMS	Configuration Management System
CO$_2$	Carbon Dioxide
CoC	Code of Conduct
CPU	Central Processing Unit
CRC	Carbon Reduction Commitment
CRT	Cathode Ray Tube
CSF	Critical Success Factor
CSI	Continual Service Improvement
CSIP	Continual Service Improvement Programme
CSO	Chief Sustainability Officer
CSR	Corporate Social Responsibility

CSV	Comma Separated Values
Defra	Department for Environment, Food and Rural Affairs
DG JRC	Directorate General Joint Research Centre
EEAS	Energy Efficiency Accreditation Scheme
EEE	Electrical and Electronic Equipment
EMS	Environmental Management System
EPA	Environmental Protection Agency
EPEAT	Electronic Product Environmental Assessment Tool
ERU	Emission Reduction Unit
EU ETS	European Union Emission Trading System
GEC	Green Electronics Council
GeSI	Global e-Sustainability Initiative
GHG	GreenHouse Gas
HFC	Hydro fluorocarbon
HHM	Half Hourly Meter
HR	Human Resources
IaaS	Infrastructure as a Service
ICT	Information Communication and Technology
IEEE	Institute of Electrical and Electronic Engineers
IP	Internet Protocol
IPCC	Intercontinental Panel on Climate Change
IPCC	Intergovernmental Panel on Climate Change
ISP	Internet Service Provider
IT	Information Technology
ITIL	IT Infrastructure Library
ITSCM	IT Service Continuity Management
ITT	Invitation to Tender
KPI	Key Performance Indicator
LCD	Liquid Crystal Display
LED	Light Emitting Diode
MFD	Multi-Functional Device
NAPM	National Association of Paper Merchants
N_2O	Nitrous Oxide

NIEA	Northern Ireland Environment Agency
OGC	Office of Government Commerce
OPEX	Operational Expenditure
PaaS	Platform as a Service
PAH	Polycyclic Aromatic Hydrocarbon
PDF	Portable Document Format
PFC	PerFluoroCarbon
PIR	Post Implementation Review
POP	Persistent Organic Pollutant
PRINCE2	PRojects IN Controlled Environment
PSU	Power Supply Unit
PUE	Power Utilisation Efficiency
PVC	Polyvinyl Chloride
QA	Quality Assurance
RACI	Responsible, Accountable, Consulted, Informed
RAM	Random Access Memory
RIO	Return on Investment
RoHS	Restriction of Hazardous Substances
ROI	Return On Investment
RTF	Rich Text Format
SaaS	Software as a Service
SACM	Service Asset & Configuration Management
SAN	Storage Area Network
SCD	Supplier and Contract Database
SEPA	Scottish Environment Protection Agency
SF$_6$	Sulphur Hexafluoride
SIP	Service Improvement Plan
SLA	Service Level Agreement
SLM	Service Level Management
SLR	Service Level Requirements
SOA	Service-Oriented Architecture
SWAC	Sea Water Air Conditioning
TBL	Triple Bottom Line

TCO	Total Cost of Ownership
tCO₂e	Tonnes of Carbon Dioxide Equivalent
UHIE	Urban Heat Island Effect
UNEP	United Nations Environment Programme
UPS	Uninterrupted Power Supply
USB	Universal Serial Bus
VGA	Video Graphics Array
VPN	Virtual Private Network
WEEE	Waste Electrical and Electronic Equipment Directive
WWF	World Wide Fund for Nature

GLOSSARY OF TERMS

Accounting   The process responsible for identifying actual costs of delivering IT services, comparing these with budgeted costs and managing variance from the budget.

Advertising Standards Authority (ASA)   UK independent watchdog committed to maintaining high standards in advertising.

Agreement   A document that describes a formal understanding between two or more parties. An agreement is not legally binding, unless it forms part of a contract.
See **Service Level Agreement, Operational Level Agreement**.

Application sizing   The activity responsible for understanding the resource requirements needed to support a new application, or a major change to an existing application. application sizing helps to ensure that the IT service can meet its agreed service level targets for capacity and performance.

Architecture   The structure of a system or IT service, including the relationships of components to each other and to the environment they are in. Architecture also includes the standards and guidelines which guide the design and evolution of the system.

Assessment   Inspection and analysis to check whether a standard or set of guidelines is being followed, that records are accurate or that efficiency and effectiveness targets are being met.
See **Audit**.

Asset   Any resource or capability. Assets of a service provider include anything that could contribute to the delivery of a service. Assets can be one of the following types: management, organisation, process, knowledge, people, information, applications, infrastructure and financial capital.

Asset management   Asset management is the process responsible for tracking and reporting the value and ownership of financial assets throughout their life cycle.

Audit   Formal inspection and verification to check whether a standard or set of guidelines is being followed, that records are accurate or that efficiency and

effectiveness targets are being met. An audit may be carried out by internal or external groups.
See **Assessment**.

Automatic Meter Reading (AMR)   Meters measure gas and electricity supply not covered by traditional HHMs.

Availability   Ability of a configuration item or IT service to perform its agreed function when required. Availability is usually calculated as a percentage.
See **Availability management**.

Availability management   A process responsible for defining, analysing, planning, measuring and improving all aspects of the availability of IT services. Availability management is responsible for ensuring that all IT infrastructure, processes, tools, roles etc. are appropriate for the agreed service level targets for availability.
See **Availability**.

Baseline   An established point for ongoing comparison.

Basel action network   A non-governmental organisation whose mission is to prevent the globalisation of the toxic chemical crisis.

Best practice   Proven activities or processes that have been successfully used by multiple organisations. ITIL is an example of best practice.

Brainstorming   A technique that helps a team to generate ideas. Ideas are usually not reviewed during the brainstorming session, but at a later stage.

Budget   A list of all the money that an organisation or business unit plans to receive, and plans to pay out, over a specified period of time.

Building management system   A computer-based control system installed in buildings that controls and monitors the building's mechanical and electrical equipment such as ventilation, lighting, power systems, fire systems and security systems.

Business case   A means of presenting information necessary to support a series of business decisions. Those decisions will (over time) increasingly commit an organisation to the achievement of the outcomes or benefits possible as a result of agreed investment in business change.

Business Relationship Management (BRM)   A formal approach to understanding, defining and supporting a broad spectrum of inter-business activities related to providing and consuming knowledge and services.

Capacity Management   The process responsible for ensuring that the capacity of IT services and the IT infrastructure is able to deliver agreed service level targets in a cost-effective and timely manner. Capacity management considers all resources required to deliver the IT service, and plans for short-, medium- and long-term business requirements.

Capacity plan   A capacity plan is used to manage the resources required to deliver IT services. The plan contains scenarios for different predictions of business demand, and costed options to deliver the agreed service-level targets.

Capacity planning   The activity within capacity management responsible for creating a capacity plan.

CApital EXPenditure (CAPEX)   The cost of purchasing something that will become a financial Asset, for example, computer equipment and buildings. The value of the asset is depreciated over multiple accounting periods.

Carbon accounting   Accounting activities undertaken to measure the amount of carbon dioxide equivalents that will not be released into the atmosphere as a result of flexible mechanisms projects created under the Kyoto Treaty.

Carbon allowances   The amount of carbon emissions an organisation is permitted under a cap-and-trade scheme.

Carbon credit   Units that can be used to finance carbon reduction schemes between trading partners and around the world. Credits can be exchanged between businesses or bought and sold in international markets at the prevailing market price.

Carbon dioxide (CO$_2$)   A colourless, odourless gas; a compound consisting of the elements carbon and oxygen. CO$_2$ is a GHG. Atmospheric CO$_2$ has increased by about 35 per cent since the beginning of the age of industrialisation.

Carbon footprint   The total set of GHG emissions caused directly and indirectly by an individual, organisation, event or product.

Carbon neutral   Achieving net zero carbon emissions by balancing a measured amount of carbon released with an equivalent amount being offset. This concept may be extended to include other GHGs measured in terms of their CO$_2$ equivalence.

Carbon offset   See **Offset** (carbon).

Carbon Quality Assurance Scheme   A scheme to advise consumers of government-approved offsets so that consumers can offset with confidence and know they are buying offsets based on accurate emission calculations and internationally approved emissions reductions.

Carbon Reduction Commitment (CRC)   A mandatory cap-and-trade scheme in the UK that will apply to large non-energy-intensive organisations in the public and private sectors. It is anticipated that the scheme will have cut carbon emissions by 1.2 million tonnes of carbon per year by 2020.

Carbon trading   Sometimes referred to as emissions trading, a market-based tool to limit GHGs. The carbon market trades emissions under cap-and-trade schemes or with credits that pay for or offset GHG reductions.

Carbon Trust   A company created by the UK government to help businesses and public organisations to reduce their emissions of CO_2 into the atmosphere, through improved energy efficiency and developing low carbon technology.

Cathode ray tube   A vacuum tube containing an electron gun and a fluorescent screen used to create images in the form of light emitted from the fluorescent screen. The image may represent electrical waveforms (oscilloscope), pictures (television, computer monitor), radar targets and others.

Certification   Issuing a certificate to confirm compliance to a standard. Certification includes a formal audit by an independent and accredited body. The term 'certification' is also used to mean awarding a certificate to verify that a person has achieved a qualification.

Certified emission reductions (CERs)   Represent carbon credits (or climate credits) issued by the Clean Development Mechanism (CDM) executive board for emission reductions achieved by CDM projects.

Chief Information Officer (CIO)   Accountable for the planning and implementation of IT systems to support cost-effectiveness, service quality and business development. Responsible for all aspects of the organisation's IT and systems.

Chief Sustainability Officer (CSO)   Accountable for the development, coordination and administration of sustainability policies and practices.

Classification   The act of assigning a category to something. Classification is used to ensure consistent management and reporting.

Clean development mechanism   An arrangement under the Kyoto Protocol allowing industrialised countries with a GHG reduction commitment to invest in projects that reduce emissions in developing countries as an alternative to more expensive emission reductions in their own countries.

Client   A generic term that means a customer, the business or a business customer. For example, client manager may be used as a synonym for account manager.

The term 'client' is also used to mean:

- a computer that is used directly by a user, for example a PC, handheld computer or workstation;
- the part of a client-server application with which the user directly interfaces, for example an email client.

Climate credit   See **Carbon credit**.

Climate change   Changes in the Earth's temperature that are often linked to rising sea levels as well as extreme weather events such as flooding and drought.

Climate Change Act   The UK's legally binding long-term framework to cut carbon emissions, the first of any country in the world.

Cloud computing   An internet-based development and use of computer in which dynamically scalable (and often virtualised) resources are provided as a service over the internet. The concept incorporates Infrastructure as a Service (IaaS), Platform as a Service (PaaS) and Software as a Service (SaaS) as well as Web 2.0.

Code of conduct   A set of rules or guidance outlining the responsibilities of or proper practices for an individual or organisation. EU examples include the Code of Conduct on Data Centres Energy Efficiency.

Code of practice   A guideline published by a public body or a standards organisation, such as ISO or BSI. Many standards consist of a code of practice and a specification. The code of practice describes recommended best practice.

Cold Cathode Fluorescent Lamp (CCFL)   Fluorescent lamps comprised of elongated internally phosphor-coated glass tubes filled with one or more gases which, when excited by an electrical signal, form plasma causing the coating to fluoresce and to illuminate the environment.

Compliance   Ensuring that a standard or set of guidelines is followed, or that proper, consistent accounting or other practices are being employed.

Component   A general term that is used to mean one part of something more complex. For example, a computer system may be a component of an IT service; an application may be a component of a release unit. Components that need to be managed should be configuration items.

Compound annual growth rate   The year-over-year growth rate of an investment over a specified period of time.

Conference call   A telephone call in which the calling party wishes to have more than one party listen in/contribute to the call.

Configuration   A generic term, used to describe a group of configuration items that work together to deliver an IT service, or a recognisable part of an IT service. Configuration is also used to describe the parameter settings for one or more CIs.

Continual Service Improvement (CSI)   CSI is responsible for managing improvements to IT service management processes and IT services. The performance of the IT service provider is continually measured and improvements are made to processes, IT services and IT infrastructure in order to increase efficiency, effectiveness and cost effectiveness.

Contract   A legally binding agreement between two or more parties.

Copenhagen Summit 2009   The 2009 United Nations Climate Change Conference, commonly known as the Copenhagen Summit, was held at the Bella Centre in Copenhagen, Denmark, between 7 December and 18 December 2009. The conference included the Fifteenth Conference of the Parties (COP 15) to the United Nations Framework Convention on Climate Change and the Fifth Meeting of the Parties (COP/MOP 5) to the Kyoto Protocol. According to the Bali Road Map, a framework for climate change mitigation beyond 2012 was to be agreed there.

Corporate Social Responsibility (CSR)   How businesses align their values and behaviour with the expectations and needs of stakeholders – not just customers and investors, but also employees, suppliers, communities, regulators, special interest groups and society as a whole. CSR describes a company's commitment to be accountable to its stakeholders.

Countermeasure   Can be used to refer to any type of control. The term countermeasure is most often used when referring to measures that increase resilience, fault tolerance or reliability of an IT service.

Course corrections   Changes made to a plan or activity that has already started, to ensure that it will meet its objectives. Course corrections are made as a result of monitoring progress.

CPU throttling   The process when the CPU tries to avoid overheating and getting damaged. If the temperature of the CPU exceeds some specified limits, the system will throttle down the CPU, allowing it to cool down and avoid damage. This process may also take place when the computer is idle. The CPU will take a lower frequency so it consumes less.

CRAMM   A methodology and tool for analysing and managing risks. CRAMM was developed by the UK government, but is now privately owned. Further information is available from http://www.cramm.com/

Critical Success Factor (CSF)   Something that must happen if a process, project, plan, or IT service is to succeed. KPIs are used to measure the achievement of each CSF. For example, a CSF of 'protect IT services when making changes' could be measured by KPIs such as 'percentage reduction of unsuccessful changes', 'percentage reduction in changes causing incidents' etc.

Culture   A set of values that is shared by a group of people, including expectations about how people should behave, ideas, beliefs and practices. See **Vision**.

Customer   Someone who buys goods or services. The customer of an IT service provider is the person or group who defines and agrees the service level targets. The term 'customer' is also sometimes informally used to mean users, for example, 'this is a customer-focused organisation'.

Data centre   A facility used to house computer systems and associated components, such as telecommunications and storage systems. Data Centres generally include redundant or back-up power supplies, redundant data communications connections, environmental controls (e.g. air conditioning, fire suppression) and security devices.

Demand management   Activities that understand and influence customer demand for services and the provision of capacity to meet these demands. At a strategic level demand management can involve analysis of patterns of business activity and user profiles. At a tactical level it can involve use of differential charging to encourage customers to use IT services at less busy times.
See **Capacity management**.

Department for Environment, Food and Rural Affairs (Defra)   A UK central government department with responsibilities that include achieving a healthy, natural environment, dealing with environmental risks and promoting a sustainable, low-carbon and resource-efficient economy.

Development   The process responsible for creating or modifying an IT service or application. Also used to mean the role or group that carries out development work.

Development environment   An environment used to create or modify IT services or applications. Development environments are not typically subjected to the same degree of control as test environments or live environments.
See **Development**.

Direct Cost   A cost of providing an IT service which can be allocated in full to a specific customer, cost centre, project etc., for example, cost of providing non-shared servers or software licences.
See **Indirect cost**.

Directorate General Joint Research Centre (DG JRC)   The Directorate General Joint Research Centre (DG JRC) of the European Commission (EC) is mandated to provide customer-driven scientific and technical support for the conception, development, implementation and monitoring of European Union (EU) policies. The JRC functions in an advisory capacity to policy-making directorates general, such as enterprise, environment and health and consumer protection, while serving also as the hub for intramural research at the EU level.

Document   Information in readable form. A document may be paper or electronic, for example, a policy statement, service level agreement, incident record, diagram of computer room layout.

Effectiveness   A measure of whether the objectives of a process, service or activity have been achieved. An effective process or activity is one that achieves its agreed objectives.
See **KPI**.

Efficiency   A measure of whether the right amount of resources have been used to deliver a process, service or activity. An efficient process achieves its objectives with the minimum amount of time, money, people or other resources. See **KPI**.

Electronic Product Environmental Assessment Tool (EPEAT)   US-based procurement tool to help evaluate, compare and select desktop computers, notebooks and monitors based on their environmental attributes.

Embodied energy   The amount of energy required to manufacture and to supply to the point of use, a service or product.

Emission reduction unit   Refers to the reduction of GHGs, where it represents one tonne of CO_2 equivalent reduced.

Emissions trading   An administrative approach used to control pollution by providing economic incentives for achieving reductions in the emissions of pollutants. An example is European Union Emission Trading System.

Energy audit   An inspection, survey and analysis of energy flows in a process or system with the objective of understanding the energy dynamics of the system under study.

Energy Act 2004   An Act to make provision for the decommissioning and cleaning up of installations and sites used for, or contaminated by, nuclear activities; to make provision about radioactive waste; to make provision for the development, regulation and encouragement of the use of renewable energy sources; to make further provision in connection with the regulation of the gas and electricity industries; to make provision for the imposition of charges in connection with the carrying out of the Secretary of State's functions relating to energy matters; to make provision for giving effect to international agreements relating to pipelines and offshore installations; and for connected purposes.

Energy Efficiency Accreditation Scheme (EEAS)   A methodology to recognise and reward achievements by organisations in reducing energy consumption. At its close in May 2008 the EEAS had accredited over 230 organisations and had been used as a model for recognising energy efficiency by organisations in the UK and as far afield as Asia and South America. In May 2008 the EEAS closed to new business and in June the Carbon Trust Standard was launched.

Energy star   An international voluntary labelling scheme for energy efficiency.

Energy Saving Trust   An independent, UK-based organisation focused on promoting action that leads to the reduction of carbon dioxide. The source of free and impartial advice and information for people across the UK looking to save energy, conserve water and reduce waste.

Environment Agency   UK government organisation responsible to the Secretary of State for Environment, Food and Rural Affairs and an assembly sponsored public body responsible to the National Assembly for Wales.

The principal aims of the agency are to protect and improve the environment and to promote sustainable development.

Environmental Management System (EMS)   Similar to other management systems that manage quality or safety, an EMS assesses an organisation's environmental strengths and weaknesses, helps to identify and manage significant impacts, measures efficiency, ensures that the organisation complies with environmental legislation and provides benchmarks for improvements.

Fair Trade   Aimed at securing better prices, decent working conditions, local sustainability and fair terms of trade for farmers and workers in the developing world. By requiring companies to pay sustainable prices (which must never fall lower than the market price), Fair Trade addresses the injustices of conventional trade, which traditionally discriminates against the poorest, weakest producers.

Fault   Synonym for error.

Flexible mechanisms   A number of mechanisms defined under the Kyoto Protocol designed to lower the overall costs of achieving emissions targets. These mechanisms enable organisations to achieve emission reductions or to remove carbon from the atmosphere cost-effectively in other countries.

Fossil fuels   Fuels formed by natural resources such as anaerobic decomposition of buried dead organisms. The age of the organisms and their resulting fossil fuels is typically millions of years, and sometimes exceeds 650 million years. These fuels contain a high percentage of carbon and hydrocarbons.

Gartner Group   A leading, global information technology research and advisory company.

Global e-Sustainability Initiative (GeSI)   Formed in 2001, GeSI promotes sustainable development in the ICT sector. It fosters global and open cooperation, informs the public of its members' voluntary actions to improve their sustainability performance and promotes technologies that foster sustainable development.

Global Compact   A United Nations initiative to encourage businesses worldwide to adopt sustainable and socially responsible policies, and to report on their implementation. The Global Compact is a principle-based framework for businesses, stating 10 principles in the areas of human rights, labour, the environment and anti-corruption.

Global warming   The progressive gradual rise of the Earth's surface temperature thought to be caused by the Greenhouse effect and responsible for changes in global climate patterns. An increase in the near surface temperature of the Earth. Global warming has occurred in the distant past as the result of natural influences, but the term is most often used to refer to the warming predicted to occur as a result of increased emissions of GHGs.

Governance   Ensuring that policies and strategy are actually implemented, and that required processes are correctly followed. Governance includes defining roles and responsibilities, measuring and reporting, and taking actions to resolve any issues identified.

GreenHouse Gases (GHGs)   Gases in an atmosphere that absorb and emit radiation within the thermal infrared range. Common GHGs include water vapour, CO_2, methane, nitrous oxide, ozone and chlorofluorocarbons.

Greenhouse Gas Protocol   A widely used international accounting tool for government and business leaders to understand, quantify and manage GHG emissions.

Green IT   The study and practice of using computing resources in an environmentally efficient way.

Green IT Champion   The person within an organisation who is responsible for delivering and owning Green IT initiatives and plans that will ultimately reduce the organisation's carbon emissions relating to IT and associated communications technology.

Green IT policy   A deliberate plan of action to guide decisions and achieve rational outcomes within the area of Green IT.

Green IT programme   The process of managing multiple interdependent projects that lead towards an improvement in an organisation's Green IT performance.

Greenpeace   A non-governmental environmental organisation, Greenpeace uses direct action, lobbying and research to achieve its goals. The global organisation does not accept funding from governments, corporations or political parties, relying on 2.86 million individual supporters and foundation grants.

Green roofing   Vegetated layers that sit on top of the conventional roof surfaces of a building.

Green wash   The practice of falsifying or exaggerating the environmental credentials of an organisation for political or financial gain or to enhance reputation.

Grid computing   The application of several computers to a single problem at the same time. Grid computing usually depends on software to divide pieces of a program amongst several computers, sometimes up to many thousands.

Guideline   A document describing best practice that recommends what should be done. Compliance to a guideline is not normally enforced.

Half Hourly Meter (HHM)   Electronic meters which record power usage in blocks of half an hour or less.

Health and Safety Executive (HSE)   The body responsible for the encouragement, regulation and enforcement of workplace health, safety and welfare, and for research into occupational risks in England, Wales and Scotland.

Hydrofluorocarbons (HFCs)   Compounds containing only hydrogen, fluorine and carbon atoms. Introduced as alternatives to ozone-depleting substances in serving many industrial, commercial and personal needs, HFCs are emitted as by-products of industrial processes and are also used in manufacturing. They do not significantly deplete the stratospheric ozone layer, but they are powerful GHGs with global warming.

IEEE 1680 Standard   A standard that provides clear and consistent performance criteria for the design of electronic products, thereby providing an opportunity to secure market recognition for efforts to reduce the environmental impact of electronic products.

Industrial Revolution   A period from the 18th to the 19th centuries in which major changes in agriculture, manufacturing, mining and transport had a profound effect on the socioeconomic and cultural conditions starting in the UK, then subsequently spreading throughout Europe, North America and eventually the world.

Information Technology (IT)   The use of technology for the storage, communication or processing of information. The technology typically includes computers, telecommunications, applications and other software. The information may include business data, voice, images, video etc. IT is often used to support business processes through IT services.

Intelligent buildings   Buildings with computer and electrical systems that sense the areas to heat and cool for maximum efficiency and then transfer air with the appropriate temperature from one place to another. Ordinary buildings often have thermostats and timers, which are often limited in that they must be set to heat or cool. Intelligent buildings have computers and sensors throughout and can not only switch from heating to cooling automatically but can also heat and cool different parts of the building simultaneously.

International Standards Organisation (ISO)   The ISO is the world's largest developer of standards. ISO is a non-governmental organisation which is a network of the national standards institutes of 156 countries. Further information about ISO is available from http://www.iso.org/

Intercontinental Panel on Climate Change (IPCC)   Established jointly by the UN Environment Programme and the World Meteorological Organization in 1988. The purpose of the IPCC is to assess information in the scientific and technical literature related to all significant components of the issue of climate change. The IPCC is also regarded as the official advisory body to the world's governments on the state of the science of climate change.

Institute of Electrical and Electronics Engineers (IEEE)   An international non-profit, professional organisation for the advancement of technology related to electricity.

Internet Service Provider (ISP)   An external service provider that provides access to the internet. Most ISPs also provide other IT services such as web hosting.

ISO 14000   A family of environmental management international standards to help organisations minimise how their operations negatively affect the environment.

IT infrastructure   All of the hardware, software, networks, facilities etc. that are required to develop, test, deliver, monitor, control or support IT services. The term 'IT infrastructure' includes all of the IT but not the associated people, processes and documentation.

IT infrastructure library   A series of books describing a best practice framework for the provision of quality IT services.

IT Service Continuity Management (ITSCM)   The process responsible for managing risks that could seriously impact IT services. ITSCM ensures that the IT service provider can always provide minimum agreed service levels, by reducing the risk to an acceptable level and planning for the recovery of IT services. ITSCM should be designed to support business continuity management.

IT Service Management (ITSM)   The implementation and management of quality IT services that meet the needs of the business. IT service management is performed by IT service providers through an appropriate mix of people, processes and IT.
See **Service Management**.

ITIL   A set of best practice guidance for IT service management. ITIL is owned by the Office of Government Commerce (OGC) and consists of a series of publications giving guidance on the provision of quality IT services, and on the processes and facilities needed to support them. See http://www.itil.co.uk/ for more information.

Key Performance Indicator (KPI)   A metric that is used to help manage a process, IT service or activity. Many metrics may be measured, but only the most important of these are defined as KPIs and used to actively manage and report on the process, IT service or activity. KPIs should be selected to ensure that efficiency, effectiveness and cost effectiveness are all managed.
See **Critical Success Factor**.

Kyoto Protocol   An international environmental Treaty produced at the UN Conference on Environment and Development (UNCED), informally known as the Earth Summit, held in Rio de Janeiro in 1992. The Treaty is intended to achieve stabilisation of GHG concentrations in the atmosphere. It establishes legally binding commitments for the reduction of greenhouse and other gases.

Landfill   Land waste disposal site in which waste is generally spread in thin layers, compacted and covered with a fresh layer of soil each day.

Legacy system   Usually an old computer system or application program that continues to be used, typically because it still functions for the users' needs, even though newer technology is available.

Life-cycle assessment The investigation and valuation of the environmental impacts of a given product or service caused or necessitated by its existence.

Life cycle The various stages in the life of an IT service, configuration item, incident, problem, change etc. The life cycle defines the categories for status and the status transitions that are permitted, for example:

- the life cycle of an application includes requirements, design, build, deploy, operate, optimise;
- the life cycle of a server may include: ordered, received, in test, live, disposed etc.

Liquid crystal display An electronically modulated optical device shaped into a thin, flat panel made up of any number of colour or monochrome pixels filled with liquid crystals and arrayed in front of a light source (backlight) or reflector. It is often utilised in battery-powered electronic devices because it uses very small amounts of electric power.

Live Refers to an IT service or configuration item that is being used to deliver service to a customer.

Live environment A controlled environment containing live configuration items used to deliver IT services to customers.

Managed services A perspective on IT services which emphasises the fact that they are managed. The term 'managed services' is also used as a synonym for out-sourced IT services.

Market space All opportunities that an IT service provider could exploit to meet the business needs of customers. The market space identifies the possible IT services that an IT service provider may wish to consider delivering.

Methane A relatively potent GHG consisting of carbon and hydrogen. Interestingly, cattle account for 16 per cent of the world's annual methane emissions into the atmosphere

Metric Something that is measured and reported to help manage a process, IT service or activity.
See **KPI**.

Mission statement A short statement that defines what an organisation is, why it exists and its reason for being.

Modelling A technique that is used to predict the future behaviour of a system, process, IT service, configuration item etc. Modelling is commonly used in financial management, capacity management and availability management.

National Association of Paper Merchants (NAPM) Accredited trade association for paper and board merchants and wholesalers.

Next Generation Networks (NGN)   Term used to describe some key architectural evolutions in telecommunication and access networks that will be deployed over the next 5–10 years. The general idea behind NGN is that one network transports all information and services (voice, data and all different media such as video) by encapsulating these into packets. NGNs are commonly built around the Internet Protocol, and therefore the term 'all-IP' is also sometimes used to describe the transformation towards NGN.

Nitrous oxide (N_2O)   A GHG. Sources of nitrous oxide include soil cultivation practices, especially the use of commercial and organic fertilizers, fossil fuel combustion, nitric acid production and biomass burning.

Northern Ireland Environment Agency (NIEA)   Responsible for implementing environmental policy and strategy in Northern Ireland and promoting key themes of sustainable development, biodiversity and climate change.

Offset (carbon)   A financial instrument representing a reduction in GHG emissions.

Open Source Software   Computer software that is available in source code form for which the source code and certain other rights normally reserved for copyright holders are provided under a software licence that permits users to study, change and improve the software.

Organisation   A company, legal entity or other institution. Examples of organisations that are not companies include the International Standards Organisation or the ITSMF. The term 'organisation' is sometimes used to refer to any entity which has people, resources and budgets, for example, a project or business unit.

Passive Infra-Red (PIR) occupancy sensing   Technology that uses sensors to turn lights on and off based on occupancy.

Percentage utilisation   The amount of time that a component is busy over a given period of time. For example, if a CPU is busy for 1,800 s in a one-hour period, its utilisation is 50 per cent

Perfluorocarbons (PFCs)   A group of human-made chemicals composed of carbon and fluorine only. These chemicals (predominantly CF_4 and C_2F_6) were introduced as alternatives, along with hydrofluorocarbons, to the ozone-depleting substances. In addition, PFCs are emitted as by-products of industrial processes and are also used in manufacturing. PFCs do not harm the stratospheric ozone layer, but they are powerful GHGs.

Persistent Organic Pollutants (POPs)   Chemical substances that persist in the environment, bio-accumulate through the food chain and pose a risk of causing adverse effects to human health and the environment.

Policy   Formally documented management expectations and intentions. Policies are used to direct decisions, and to ensure consistent and appropriate development and implementation of processes, standards, roles, activities, IT infrastructure etc.

Polycyclic Aromatic Hydrocarbons (PAHs)   A diverse class of organic compounds. There are several hundred PAHs, which usually occur as complex mixtures rather than as individual compounds. The most well-known PAH is benzo[a]pyrene (BaP). PAHs are flammable, colourless solids or crystals at room temperature with no perceptible odour.

Post Implementation Review (PIR)   A review that takes place after a change or a project has been implemented. A PIR determines if the change or project was successful, and identifies opportunities for improvement.

Practice   A way of working or a way in which work must be done. Practices can include activities, processes, functions, standards and guidelines.
See **Best practice**.

Procedure   A document containing steps that specify how to achieve an activity. Procedures are defined as part of processes.
See **Work instruction**.

Production environment   Synonym for live environment.

Programme   A number of projects and activities that are planned and managed together to achieve an overall set of related objectives and other outcomes.

Project   A temporary organisation, with people and other assets required to achieve an objective or other outcome. Each project has a life cycle that typically includes initiation, planning, execution, closure etc. Projects are usually managed using a formal methodology such as PRINCE2.

Quality   The ability of a product, service or process to provide the intended value. For example, a hardware component can be considered to be of high quality if it performs as expected and delivers the required reliability. Process quality also requires an ability to monitor effectiveness and efficiency, and to improve them if necessary.
See **Quality Management System**.

Quality Assurance (QA)   The process responsible for ensuring that the quality of a product, service or process will provide its intended value.

Quick win   An improvement activity which is expected to provide a return on investment in a short period of time with relatively small cost and effort.

RACI matrix   An acronym for Responsible, Accountable, Consulted and Informed. A RACI matrix helps to identify all the activities or decision-making authorities undertaken in an organisation and sets them against individuals or roles. At each intersection of activity and role, it is possible to assign somebody responsible, accountable, consulted or informed for that activity or decision.

Renewable energy Energy which comes from natural resources such as sunlight, wind, rain, tides and geothermal heat that is naturally replenished.

Retire Permanent removal of an IT service, or other configuration item, from the live environment. Retired is a stage in the life cycle of many configuration items.

Return On Investment (ROI) A measurement of the expected benefit of an investment. In the simplest sense it is the net profit of an investment divided by the net worth of the assets invested.

Rights Entitlements, or permissions, granted to a user or role, for example the right to modify particular data, or to authorise a change.

Rio Declaration on Environment and Development Generally shortened to the 'Rio Declaration'. A document produced at the 1992 UN Conference on Environment and Development normally referred to as the Earth Summit. The Rio Declaration consisted of 27 principles intended to guide future sustainable development worldwide, reaffirming the Declaration of the United Nations Conference on the Human Environment, adopted at Stockholm on 16 June 1972, and seeking to build upon it.

Risk A possible event that could cause harm or loss, or affect the ability to achieve objectives. A risk is measured by the probability of a threat, the vulnerability of the asset to that threat and the impact it would have if it occurred.

Risk assessment The initial steps of risk management. Analysing the value of assets to the business, identifying threats to those assets and evaluating how vulnerable each asset is to those threats. Risk assessment can be quantitative (based on numerical data) or qualitative.

Risk management The process responsible for identifying, assessing and controlling risks.
See **Risk assessment**.

Running costs Synonym for operational costs.

Sarbanes Oxley Act (SOX) US law that came into force in July 2002 and introduced major changes to the regulation of corporate governance and financial practice. It is named after Senator Paul Sarbanes and Representative Michael Oxley, who were its main architects, and it set a number of non-negotiable deadlines for compliance.

Scottish Environment Protection Agency Responsible for implementing environmental policy and strategy in Scotland and promoting key themes of sustainable development, biodiversity and climate change.

Sea Water Air Conditioning (SWAC)   A method for cooling data centres and other buildings by drawing sea water from a depth of approximately 1,600 feet (500 m) to a cooling station onshore, where heat exchangers enable it to circulate fresh water in a closed loop to buildings. After passing through the heat exchangers, the warmed sea water will be returned to the ocean at a shallower depth, using a diffuser that ensures proper mixing and dilution.

Server   A computer that is connected to a network and provides software functions that are used by other computers.

Service   A means of delivering value to customers by facilitating outcomes that customers want to achieve without the ownership of specific costs and risks.

Service Improvement Plan (SIP)   A formal plan to implement improvements to a process or IT service.

Service Level Agreements (SLAs)   Written agreement between an IT service provider and customer defining key service level targets and the responsibilities of each party.

Service Level Management (SLM)   The process responsible for negotiating service level agreements, and ensuring that these are met. SLM is responsible for ensuring that all IT service management processes, operational level agreements and underpinning contracts are appropriate for the agreed service level targets. SLM monitors and reports on service levels, and holds regular customer reviews.

Service management   Service management is a set of specialised organisational capabilities for providing value to customers in the form of services.

Service management life cycle   An approach to IT service management that emphasises the importance of coordination and control across the various functions, processes and systems necessary to manage the full life cycle of IT services. The service management life-cycle approach considers the strategy, design, transition, operation and continuous improvement of IT services.

Service Oriented Architecture (SOA)   A flexible set of design principles used during the phases of systems development and integration. A deployed SOA-based architecture will provide a loosely integrated suite of services that can be used within multiple business domains.

Silicon trading   A term used to describe the practice of growing the carbon footprint of ICT to reduce the overall carbon footprint of the organisation.

Small-to-Medium Enterprise (SME)   A term used to describe organisations with between approximately 50 and 249 employees. Despite governments and many of the multinational organisations targeting this group for special financial business support, there is no single definition for a SME.

SMART 2020   A study commissioned by The Climate Group that stated that ICT could drive efficiency across the economy and deliver emission savings of

15% – 7.8 GtCO₂e – of global BAU emissions by 2020. SMART is an acronym for Standards, Monitoring, Rethink, Transform.

Specification   A formal definition of requirements. A specification may be used to define technical or operational requirements, and may be internal or external. Many public standards consist of a code of practice and a specification. The specification defines the standard against which an organisation can be audited.

Stakeholder   All people who have an interest in an organisation, project, IT service etc. Stakeholders may be interested in the activities, targets, resources or deliverables. Stakeholders may include customers, partners, employees, shareholders, owners etc.

Storage Area Network (SAN)   Technology that allows multiple servers to connect to a centralised pool of disk storage.

Storage management   The process responsible for managing the storage and maintenance of data throughout its life cycle.

Strategic   The highest of three levels of planning and delivery (strategic, tactical, operational). Strategic activities include objective setting and long-term planning to achieve the overall vision.

Strategy   A strategic plan designed to achieve defined objectives.

Sulphur hexafluoride   A compound composed of one sulphur and two oxygen molecules. Sulphur dioxide emitted into the atmosphere through natural and anthropogenic processes is changed in a complex series of chemical reactions in the atmosphere to sulphate aerosols.

Supplier   A third party responsible for supplying goods or services that are required to deliver IT services. Examples of suppliers include commodity hardware and software vendors, network and telecom providers and outsourcing organisations.
See **Supply chain**.

Supplier and Contract Database (SCD)   A database or structured document used to manage supplier contracts throughout their life cycle. The SCD contains key attributes of all contracts with suppliers, and should be part of the service knowledge management system.

Supplier management   The process responsible for ensuring that all contracts with suppliers support the needs of the business, and that all suppliers meet their contractual commitments.

Supply chain   The activities in a value chain carried out by suppliers. A supply chain typically involves multiple suppliers, each adding value to the product or service.

Sustainability   The ability to maintain a certain process or state.

Sustainable urban drainage system   A sequence of water management practices and facilities designed to drain surface water in a manner that will provide a more sustainable approach than what has been the conventional practice of routing run-off through a pipe to a watercourse.

TBL accounting   A term used to refer to an organisation reporting on the financial, environmental and social returns or impacts of its investments. Such reports provide a picture of the long-term stability of an enterprise in terms of its economic vitality, social relationships with stakeholders, and environmental compliance and integrity. TBL accounting can assist businesses and their stakeholders to evaluate impact on different dimensions of sustainable development.

Teleconferencing   Teleconference is the live exchange of information between people remote from one another but linked by a telecommunications system.

Thick client   A client computer which does as much processing as possible and passes only data for communications and storage to the server.

Thin client   A client computer which depends primarily on the central server for processing activities, and which mainly focuses on conveying input and output between the user and the remote server.

Throughput   A measure of the number of transactions, or other operations, performed in a fixed time, for example, 5,000 emails sent per hour, or 200 disk I/Os per second.

Total Cost of Ownership (TCO)   A methodology used to help make investment decisions. TCO assesses the full life-cycle cost of owning a configuration item, not just the initial cost or purchase price.

Transaction   A discrete function performed by an IT service, for example transferring money from one bank account to another. A single transaction may involve numerous additions, deletions and modifications of data. Either all of these complete successfully or none of them is carried out.

Transition   A change in state, corresponding to a movement of an IT service or other configuration item from one life-cycle status to the next.

Trojan infections   Malicious programs that can cause damage to your computer. The name derives from the horse of the same name, as it appears as something different from what it actually is.

Trend analysis   Analysis of data to identify time-related patterns.

United Nations   An international organisation whose stated aims include facilitating cooperation in international law, international security, economic development and social progress.

United Nations Environment Programme   The UN system's designated entity for addressing environmental issues at the global and regional level. Its mandate is to coordinate the development of environmental policy consensus by keeping the global environment under review and bringing emerging issues to the attention of governments and the international community for action.

Urban Heat Island Effect (UHIE)   The effect of a built-up area which is significantly warmer than its surroundings.

US Environmental Protection Agency (EPA)   An agency of the government of the United States charged to protect human health and the environment, by writing and enforcing regulations based on laws passed by US Congress. The EPA was proposed by President Richard Nixon and began operation on 2 December 1970, when its establishment was passed by Congress, and signed into law by President Nixon, and has since been chiefly responsible for the environmental policy of the United States.

Videoconferencing   A set of interactive telecommunication technologies which allow two or more locations to interact via two-way video and audio transmissions simultaneously.

Virtualisation   The abstraction of computer resources. In case of server consolidation, many small physical servers are replaced by one larger physical server, to increase the utilisation of costly hardware resources such as the CPU.

Vision   A description of what the organisation intends to become in the future. A vision is created by senior management and is used to help influence culture and strategic planning.

Voiceconferencing   See **Teleconferencing**.

Voluntary Emission Reduction (VER)   An emission reduction that has been achieved outside of compulsion. VERs are carbon credits developed by carbon offset providers which are not certified.

Waste Electrical and Electronic Equipment Directive   An EU directive that imposes the responsibility for the disposal of waste electrical and electronic equipment on the manufacturers of such equipment.

Web 2.0   A term describing a second generation of web development and design including the development and evolution of web-based communities, hosted services and applications, such as social-networking sites, video-sharing sites, wikis and blogs.

Windows Active Directory   Developed by Microsoft, Windows Active Directory allows administrators to assign policies, deploy software and apply critical updates to an organisation. Active Directory stores information and settings in a central database and can operate on a small installation with a few computers, users and printers to tens of thousands of users, many different domains and large server farms spanning many geographical locations.

Windows Task Scheduler   Developed by Microsoft, Task Scheduler is a component of Microsoft Windows that provides the ability to schedule the launch of programs or scripts at predefined times or after specified time intervals.

Wireless connectivity   The transfer of information over a distance without the use of electrical conductors or wires. The distances involved may be short (a few metres as in television remote control) or long (thousands or millions of kilometers for radio communications). When the context is clear, the term is often shortened to 'wireless'. Wireless communication is generally considered to be a branch of telecommunications.

World Wide Fund for Nature   Launched on 23 November 1961, a globally recognised charitable organisation, dedicated to addressing issues relating to species and habitat survival and other environmental issues.

World Wide Web (WWW)   Commonly known as the Web, it is a system of interlinked hypertext documents contained on the internet.

NOTES

1. In January 2010 the UN climate science panel was accused of wrongly linking global warming to an increase in the number and severity of natural disasters such as hurricanes and floods. In a report by *The Sunday Times* on 24 January 2010, the Intergovernmental Panel on Climate Change (IPCC) was accused of basing its claims on an unpublished report that had not been subjected to routine scientific scrutiny, and ignored warnings from scientific advisers that the evidence supporting the link was too weak. In reply, the IPCC released a press statement accusing *The Sunday Times* of running a misleading and baseless story attacking the way in which the Fourth Assessment Report of the IPCC handled an important question concerning recent trends in economic losses from climate-related disasters. It said that *The Sunday Times* article got the story wrong on two key points. The first was that the Report incorrectly assumed that a brief section on trends in economic losses from climate-related disasters was all that the IPCC Fourth Assessment Report (2007) had to say about changes in extremes and disasters, when, in fact, the Fourth Assessment Report reaches many important conclusions, at many locations in the report, about the role of climate change in extreme events. Second, the IPCC felt the problem with the article in *The Sunday Times* was its 'baseless attack on the section of the report on trends in economic losses from disasters'. The IPCC felt that this particular section of the Report was a balanced treatment of a complicated and important issue, and that it clearly made the point that one study detected an increase in economic losses, corrected for values at risk, but that other studies had not detected such a trend.

2. The Carbon Trust is a not-for-profit company with the mission to accelerate the move to a low-carbon economy. It provides specialist support to help businesses and the public sector cut carbon emissions, save energy and commercialise low-carbon technologies. To learn more about the Carbon Trust, visit their website at http://www.carbontrust.co.uk

3. Fairtrade is a scheme that focuses on responsible commercial working practices, from producers and farmers all the way through to the consumer. By requiring organisations to pay sustainable prices (which must never fall lower than the market price), Fairtrade addresses the injustices of some conventional trading practices, which traditionally discriminate against the poorest and weakest producers. It enables them to improve their position and have more control over their lives.

4. The Carbon Trust has a series of posters on its website that are free to download to anyone who wishes to register with them.

5. In 2009, 10 mobile phone companies including Apple, LG, Motorola, Nokia and Sony Ericsson agreed to manufacture phones designed to use a universal charger based on a micro-USB connector. Discussions between the phone companies and EU commission officials produced a 'Memorandum of Understanding' indicating that the first generation of 'inter-chargeable' mobile phones will reach the EU market from 2010 onwards.

6. http://www.timesonline.co.uk/tol/news/environment/article6973577.ece

7. Adapted from an article posted on the WWF-UK website www.wwf.org.uk/news

8. http://www.theclimategroup.org/our-news/news/2008/6/19/smart-2020-enabling-the-low-carbon-economy-in-the-information-age/

9. http://www.unglobalcompact.org

10. Information reproduced and adapted with the permission of Gary Mills of Fusion Group.

11. Source: EU2 Analysis and Market Survey for European Building Technologies in Central and Eastern European Countries – GOPA.

12. The ISO 14000 family addresses various aspects of environmental management. The very first two standards, ISO 14001:2004 and ISO 14004:2004, deal with Environmental Management Systems (EMSs). ISO 14001:2004 provides the requirements for an EMS and ISO 14004:2004 gives general EMS guidelines. The other standards and guidelines in the family address specific environmental aspects, including: labelling, performance evaluation, life-cycle analysis, communication and auditing.

13. http://business.timesonline.co.uk/tol/business/related_reports/best_green_companies/

14. Lewis Carroll, famous English author, mathematician, logician, Anglican deacon and photographer.

15. You can read the Act in full on the Office of Public Sector Information (OPSI): UK Statute Law Database – The Climate Change Act 2008 web page.

16. AMR meters have been developed for gas and electricity so that consumers can access data on supplies. These meters provide consumers with access to their energy supply data. There is a wide range of AMR equipment available; however, CRC will only capture AMR meters which can be read remotely.

17. The Carbon Trust Standard certifies that an organisation has genuinely reduced its carbon footprint and is committed to making further reductions

year-on-year. Assessment against the Standard is undertaken by independent third-party assessors, based on the evidence provided by the participating organisation.

18. Information adapted from the carbon reduction commitment user guide issued by the Department of Energy and Climate Change.

19. Adapted from an article written for www.climatechangecorp.com by Dr Michael Gell of Zanfeon Ltd. For regular updates to the CRC, please refer to the UK Department of Energy and Climate Change.

20. Diagram reproduced with the permission of the Sustain IT UK Centre for Economic and Environmental Development.

21. Gartner Group.

22. Environment Agency – Department for Business Enterprise and Regulatory Reform – WEEE Business User Fact Sheet.

23. Duty of Care – Section 34 of the EPA90 Act imposes a duty of care on those concerned with the controlled waste. This applies to those who produce, import, carry, treat or dispose of controlled waste. The legal definition of waste is 'any substance or object which the producer or the person in possession of it discards or intends or is required to discard'. Waste is therefore anything you own, or your business produces, that you want to get rid of. It can be household, commercial or industrial waste. Certain wastes are not included within the definition; these are agricultural wastes, wastes from mines and quarries and certain radioactive wastes.
Special Waste Regulations – These regulations cover the disposal, carrying or receiving of special wastes. Guidance is available on what constitutes a special waste but in general it covers hazardous and toxic wastes, for example acids, industrial solvents, pharmaceutical compounds, waste oils and wood preservatives.

24. Exporting Harm: The High-Tech Trashing of Asia (pdf). Basel Action Network. http://www.ban.org/E-waste/technotrashfinalcomp.pdf

25. For the full 2007 EPEAT Environmental Benefits Report, see http://www.epeat.net/Docs/EnvironmentalBenefits2007.pdf

26. The GEC is a program of the International Sustainable Development Foundation which is a not-for-profit organisation located in Portland, Oregon, USA. The GEC was established in 2006 with a mission to inspire and support the effective design, manufacture, use and recovery of electronic products to contribute to a healthy, fair and prosperous world. Through its partnerships with the electronics industry and environmental organisations, government agencies, manufacturers and other interested stakeholders, the GEC implements market-driven systems to recognise and reward environmentally preferable electronic products and builds the capacity of individuals and organisations to design and manage the life cycle of

electronic products to improve their environmental and social performance. For more information, visit www.greenelectronicscouncil.org

27. http://www.computerweekly.com/Articles/2008/09/30/232493/councils-network-exposed-after-server-sold-on-ebay.htm

28. http://www.greenroofs.com/projects/pview.php?id=21

29. Information provided by Recycled Paper Supplies http://rps.gn.apc.org/index.htm

30. Source: publishing house Piper Jaffray & Co; US investment bank.

31. The Energy Saving Trust is an independent, UK-based organisation focused on promoting action that leads to the reduction of CO_2 emissions. They are a source of free advice and information for people across the UK looking to save energy, conserve water and reduce waste. For more information, visit their website at http://www.energysavingtrust.org.uk

32. Tests performed and published by CNet.com, http://reviews.cnet.com/green-tech/monitor-power-efficiency/?tag=greenGuideBodyColumn.1

33. Computer Aid International is a UK-registered charity that aims to reduce poverty through practical ICT solutions. Computer Aid provides high-quality, professionally refurbished computers for reuse in education, health and not-for-profit organisations in developing countries. Computer Aid has provided over 150,000 PCs to more than 100 countries across Africa and South America, making them the world's largest ICT for Development provider. To learn more about Computer Aid, visit their website at http://www.computeraid.org

34. Electricity Consumption and Efficiency Trends in European Union – Status Report 2009 European Commission Joint Research Centre Institute for Energy.

35. Corin Ltd are a part of the Corin Group who develop, produce and manufacture reconstructive orthopaedic devices.

36. Spotify is a registered trademark of Spotify Ltd.

37. The Data Protection Act 1998 requires anyone who handles personal information to comply with a number of important principles. It also gives individuals rights over their personal information.

38. The Office of Government and Commerce (OGC) is an independent office of Her Majesty's Treasury, a department of state in UK government.

39. Service Portfolio describes a service provider's services across its conceptual or pipeline services, its operations services within the service catalogue and retired services.

40. An SLR is a customer requirement for an aspect of a service. It is based on business objectives and is used to negotiate service level targets as part of an SLA.

41. A full explanation of business risk and risk management is provided as part of the conclusion to this book.

42. A CI is any component of a service which we wish to record, maintain attribute information, manage and control through Change Management.

43. A CMS is a supporting system holding information on assets and CIs which are in the scope of SACM.

44. Organisational Change Management is addressed in Chapter 12.

45. The CAB is a group of advisers or parties interested in a change. It may have fixed or floating membership.

46. A function is a role or group of people with responsibility for specific outcomes, needing specific resources and capabilities for the purpose.

47. This is a centralised location for the monitoring of services ensuring timely response to events.

48. PUE is the ratio of power delivered to IT equipment to the total amount of power used by the data centre facility to support it, i.e. in cooling or power distribution.

49. W. Edwards Deming was a well-known management theorist. His theories were designed to lead to higher quality and productivity as well as enhanced competitive position.

50. UNEP is the United Nations system's designated entity for addressing environmental issues at a global and regional level. Its mandate is to coordinate the development of environmental policy consensus by keeping the global environment under review and bringing emerging issues to the attention of governments and the international community for action.

51. To find out more about the Global Action Plan, see their website: http://www.globalactionplan.org.uk/

52. PRINCE2 (PRojects IN Controlled Environments) is a process-based method for effective project management. PRINCE2 is a de facto standard used extensively by the UK Government and is widely recognised and used in the private sector, both in the UK and internationally. The method PRINCE2 is in the public domain, offering non-proprietarily best practice guidance on project management. PRINCE2 is a registered trademark of OGC.

53. To learn more about Professor JP Kotter, see: http://www.kotterinternational.com

INDEX

bcs

The
Chartered
Institute
for IT

Enabling the
information society

PROMOTE YOURSELF WITH THE GREEN IT FOUNDATION CERTIFICATE

- Prove you can create a Green IT strategy
- Show that you're aware of carbon energy accounting
- Demonstrate that you understand the regulations, legislation and policies within the green environment

To demonstrate that you fully understand the fundamental principles of Green IT, the new Foundation Certificate from BCS, The Chartered Institute for IT, will develop your knowledge of Green IT and provide you with a world class qualification.

Promote yourself and demonstrate your environmental credentials to your organisation by being one of the first to gain this world class qualification.

Visit **www.bcs.org/greenit/promo** to enable you to take the on-line Foundation Certificate in Green IT through a special promotional offer.

BUSINESS ANALYSIS
SECOND EDITION

James Cadle, Malcolm Eva, Keith Hindle, Debra Paul, Craig Rollaston, Dot Tudor, Donald Yeates

Provides workable skills and techniques, underpinned with academic theory. This practical introductory guide is suitable for those involved with various aspects of business analysis or improving the effectiveness of IT and its alignment with business objectives.

- **New edition of bestselling title**

- **A practical introduction for anyone involved in business analysis, improving efficiency or aligning IT with business objectives**

- **Key areas include: practical business analysis techniques, systems development, process management and resource management**

- **'Great breadth, good depth'**
  (Neil Venn, Seabright Consulting)

AUTHOR INFORMATION
Business Analysis has been written by a team of experts who are practitioners and educators in the business analysis field.

www.bcs.org/books/businessanalysis2

ISBN:	978-1-906124-61-8
Format:	Paperback, 246 x 172, 256pp
Price:	£29.95
Published:	July 2010

BCS Books, Turpin Distribution, Pegasus Drive, Stratton Business Park, Biggleswade, Bedfordshire, SG18 8TQ, UK
Tel +44 (0)1767 604951 custserv@turpin-distribution.com

A PRAGMATIC GUIDE TO
BUSINESS PROCESS MODELLING
SECOND EDITION

Jon Holt

A practical handbook for carrying out accurate and
effective process modelling, drawn from the author's
considerable experience in consulting.

- **Bestselling business process modelling title, now in
  its 2nd edition**

- **Completely revised and expanded to include 5 new
  chapters covering presentation of process information,
  a teaching guide, enterprise architecture and more**

AUTHOR INFORMATION
Jon Holt is an award-winning author and public speaker,
specialising in all aspects of systems, process and
competency modelling. Jon's other work interests include
enterprise architecture, standards and education, and he
has previously held a variety of positions at universities in
the UK and USA. Jon is a Fellow of the IET and BCS.

www.bcs.org/books/processmodelling

ISBN:	978-1-906124-12-0
Format:	Paperback, 246 x 172, 248pp
Price:	£29.95
Published:	July 2009

BCS Books, Turpin Distribution, Pegasus Drive, Stratton
Business Park, Biggleswade, Bedfordshire, SG18 8TQ, UK
Tel +44 (0)1767 604951 custserv@turpin-distribution.com